COMPREHENSIVE BIOCHEMISTRY

ELSEVIER PUBLISHING COMPANY

335 Jan van Galenstraat, P.O. Box 211, Amsterdam

AMERICAN ELSEVIER PUBLISHING COMPANY, INC.

52, Vanderbilt Avenue, New York, N.Y. 10017

ELSEVIER PUBLISHING COMPANY LIMITED

Rippleside Commercial Estate, Barking, Essex

Library of Congress Catalog Card Number 62–10359

With 51 illustrations and 11 tables

PRINTED IN THE NETHERLANDS

COMPREHENSIVE BIOCHEMISTRY

COMPREHENSIVE
BIOCHEMISTRY

SECTION I (VOLUMES 1–4)
PHYSICO-CHEMICAL AND ORGANIC ASPECTS
OF BIOCHEMISTRY

SECTION II (VOLUMES 5–11)
CHEMISTRY OF BIOLOGICAL COMPOUNDS

SECTION III (VOLUMES 12–16)
BIOCHEMICAL REACTION MECHANISMS

SECTION IV (VOLUMES 17–21)
METABOLISM

SECTION V (VOLUMES 22–29)
CHEMICAL BIOLOGY

HISTORY OF BIOCHEMISTRY (VOLUME 30)
GENERAL INDEX (VOLUME 31)

COMPREHENSIVE
BIOCHEMISTRY

EDITED BY

MARCEL FLORKIN

Professor of Biochemistry, University of Liège (Belgium)

AND

ELMER H. STOTZ

*Professor of Biochemistry, University of Rochester, School of Medicine
and Dentistry, Rochester, N.Y. (U.S.A.)*

VOLUME 22

BIOENERGETICS

ELSEVIER PUBLISHING COMPANY

AMSTERDAM · LONDON · NEW YORK

1967

CONTRIBUTORS TO THIS VOLUME

F. J. BULLOCK, B.S., A.M., PH.D.

Senior Chemist, Organic and Medicinal Chemistry Group, Arthur D. Little, Inc., Acorn Park, Cambridge, Mass. 02140 (U.S.A.)

TH. FÖRSTER, DR. PHIL. NAT.

Professor of Physiological Chemistry, Fakultät für Natur- und Geisteswissenschaften, Technische Hochschule Stuttgart, Wiederholdstrasse 15, 7 Stuttgart-N. (Deutschland)

PETER MITCHELL, PH.D.

Glynn Research Laboratories, Bodmin, Cornwall (Great Britain)

ALBERTE PULLMAN, D.Sc.

Director of Research at the National Research Council

and

BERNARD PULLMAN, D.Sc.

Professor of Quantum Chemistry at the Sorbonne; Director, Institut de Biologie Physico-Chimique, 13, rue Pierre Curie, Paris 5e (France)

GENERAL PREFACE

The Editors are keenly aware that the literature of Biochemistry is already very large, in fact so widespread that it is increasingly difficult to assemble the most pertinent material in a given area. Beyond the ordinary textbook the subject matter of the rapidly expanding knowledge of biochemistry is spread among innumerable journals, monographs, and series of reviews. The Editors believe that there is a real place for an advanced treatise in biochemistry which assembles the principal areas of the subject in a single set of books.

It would be ideal if an individual or small group of biochemists could produce such an advanced treatise, and within the time to keep reasonably abreast of rapid advances, but this is at least difficult if not impossible. Instead, the Editors with the advice of the Advisory Board, have assembled what they consider the best possible sequence of chapters written by competent authors; they must take the responsibility for inevitable gaps of subject matter and duplication which may result from this procedure.

Most evident to the modern biochemist, apart from the body of knowledge of the chemistry and metabolism of biological substances, is the extent to which he must draw from recent concepts of physical and organic chemistry, and in turn project into the vast field of biology. Thus in the organization of Comprehensive Biochemistry, the middle three sections, Chemistry of Biological Compounds, Biochemical Reaction Mechanisms, and Metabolism may be considered classical biochemistry, while the first and last sections provide selected material on the origins and projections of the subject.

It is hoped that sub-division of the sections into bound volumes will not only be convenient, but will find favour among students concerned with specialized areas, and will permit easier future revisions of the individual volumes. Toward the latter end particularly, the Editors will welcome all comments in their effort to produce a useful and efficient source of biochemical knowledge.

M. FLORKIN

E. H. STOTZ

Liège/Rochester

PREFACE TO SECTION V

(VOLUMES 22–29)

After Section IV (*Metabolism*), Section V is devoted to a number of topics which, in an earlier stage of development, were primarily descriptive and included in the field of Biology, but which have been rapidly brought to study at the molecular level. "*Comprehensive Biochemistry*", with its chemical approach to the understanding of the phenomena of life, started with a first section devoted to certain aspects of organic and physical chemistry, aspects considered pertinent to the interpretation of biochemical techniques and to the chemistry of biological compounds and mechanisms. Section II has dealt with the organic and physical chemistry of the major organic constituents of living material, including a treatment of the important biological high polymers, and including sections on their shape and physical properties. Section III is devoted primarily to selected examples from modern enzymology in which advances in reaction mechanisms have been accomplished. After the treatment of Metabolism in the volumes of Section IV, "*Comprehensive Biochemistry*", in Section V, projects into the vast fields of Biology and deals with a number of aspects which have been attacked by biochemists and biophysicists in their endeavour to bring the whole field of life to a molecular level. Besides the chapters often grouped under the heading of molecular biology, Section V also deals with modern aspects of bioenergetics, immuno-chemistry, photobiology and finally reaches a consideration of the molecular phenomena that underlie the evolution of organisms.

Liège/Rochester
February 1967

M. FLORKIN
E. H. STOTZ

CONTENTS

VOLUME 22

BIOENERGETICS

General Preface. vii
Preface to Section V viii

Chapter I. Quantum Biochemistry

by ALBERTE PULLMAN AND BERNARD PULLMAN

1. Introduction. 1
    a. The methods and the scope of quantum biochemistry. 3
    b. Present status of the calculations. 15
2. Aspects of the electronic structure of the nucleic acids 17
    a. Overall results. 17
    b. The significance of resonance energy 21
    c. Van der Waals–London interactions 23
    d. Chemical reactivity. 26
3. Electron-donor and -acceptor properties of biomolecules and charge-transfer
    complexes. 27
4 The structure of proteins (and their constituents) and the problem of semiconduc-
    tivity in biopolymers 35
5. The origin of the "energy wealth" in the energy-rich phosphates 41
6. The mechanism of enzyme and coenzyme activity. 43
7. The mechanism of chemical carcinogenesis 49
    a. Structure–activity correlations in aromatic hydrocarbons 49
    b. Interactions between the carcinogens and possible cellular receptors 53
8. Conclusions 55

References 56

Chapter II. Mechanisms of Energy Transfer

by TH. FÖRSTER

1. Introduction. 61
2. Theory of excitation transfer. 63
    a. Classifications. 63
    b. Strong coupling 64
    c. Weak coupling 65

    d. Very weak coupling 67
    e. Long range dipole–dipole transfer 68
  3. Experimental investigations 71
    a. Methods 71
    b. Delocalization 71
    c. Single-step transfer 72
    d. Multi-step transfer 74
  4. Applications to biological systems 76

References 78

Chapter III. Charge Transfer in Biology
Section a. Donor–Acceptor Complexes in Solution

by F. J. BULLOCK

  1. Introduction 81
  2. Energetics 82
  3. Characteristics of the charge-transfer absorption band 89
  4. Paramagnetic complexes and electron-transfer reactions 93
  5. Equilibrium constants 99
  6. Contact charge transfer 100
  7. Solvent effects 101
  8. Other useful physical methods for the study of complex formation 102
    a. Infrared spectroscopy 102
    b. Nuclear magnetic resonance spectroscopy 103
    c. Polarography 103
    d. Fluorescence and phosphorescence 105
    e. Temperature-jump relaxation and flash techniques 106
  9. Effect of complex formation on reaction rates 106
 10. Complexes of biological materials 109
    a. Flavins 109
    b. Pyridine nucleotides 135
    c. Other complexes 138
 11. Complexes of metal cations 140

References 143

Chapter III. Charge Transfer in Biology
Section b. Transfer of Charge in the Organic Solid State

by F. J. BULLOCK

  1. Transfer of charge by semiconduction 150
  2. Photoconductivity 154
  3. Donor–acceptor complexes 161

4. Models of lamellar systems 162
5. Dye-sensitized photoconductivity 163

References 164

Chapter IV. *Active Transport and Ion Accumulation*

by PETER MITCHELL

1. Introduction 167
  a. Some relationships between chemical reaction and transport 167
  b. Chemiosmotic processes 170
2. Translocation catalysis 172
  a. The facilitation of solute diffusion by catalytic carriers 172
  b. Group translocation 174
  c. Classification of translocation reactions 175
3. Translocation catalysis through lipoprotein membranes 176
4. The general mechanisms of translocation catalysis 178
  a. Mobile *versus* fixed carriers 178
  b. The carrier centre 178
5. Secondary translocation 179
  a. Non-coupled solute translocation: uniport 179
    (*i*) Circulating carrier uniporters, 179 – (*ii*) "Single channel" or "pore"
    uniporters, 180 –
  b. Anti-coupled solute translocation: antiport 180
  c. Sym-coupled solute translocation: symport 181
    (*i*) The Na^+–glucose symporter, 181 – (*ii*) Na^+–amino acid symporters, 182 –
  d. Proton-coupled solute translocation 184
    (*i*) The ATP/ADP translocator of mitochondria, 185 – (*ii*) Proton- or hydroxyl
    ion-coupled anion translocators, 185 –
6. Primary translocation 186
  a. The Na^+/K^+-antiporter-ATPase 186
  b. The H^+-translocator-ATPase 187
  c. The H^+-translocator oxido–reductases 189
7. The coupling of primary and secondary translocation 190
8. Postcript 190

Acknowledgements 191

References 192

Subject Index 199

COMPREHENSIVE BIOCHEMISTRY

Section I — Physico-Chemical and Organic Aspects of Biochemistry
Volume 1. Atomic and molecular structure
Volume 2. Organic and physical chemistry
Volume 3. Methods for the study of molecules
Volume 4. Separation methods

Section II — Chemistry of Biological Compounds
Volume 5. Carbohydrates
Volume 6. Lipids — Amino acids and related compounds
Volume 7. Proteins (Part 1)
Volume 8. Proteins (Part 2) and Nucleic acids
Volume 9. Pyrrole pigments, isoprenoid compounds, phenolic plant constituents
Volume 10. Sterols, bile acids and steroids
Volume 11. Water-soluble vitamins, hormones, antibiotics

Section III — Biochemical Reaction Mechanisms
Volume 12. Enzymes — general considerations
Volume 13 (second revised edition). Enzyme nomenclature
Volume 14. Biological oxidations
Volume 15. Group-transfer reactions
Volume 16. Hydrolytic reactions; cobamide and biotin coenzymes

Section IV — Metabolism
Volume 17. Carbohydrate metabolism
Volume 18. Lipid metabolism
Volume 19. Metabolism of amino acids, proteins, purines, and pyrimidines
Volume 20. Metabolism of porphyrins, steroids, isoprenoids, flavonoids and fungal substances
Volume 21. Vitamins and inorganic metabolism

Section V — Chemical Biology
Volume 22. Bioenergetics
Volume 23. Cytochemistry
Volume 24. Biological information transfer. Viruses. Chemical immunology
Volume 25. Regulatory functions, membrane phenomena
Volume 26. Extracellular and supporting structures
Volume 27. Photobiology, ionizing radiations
Volume 28. Morphogenesis, differentiation and development
Volume 29. Comparative biochemistry, molecular evolution

Volume 30. History of biochemistry

Volume 31. General index

Chapter I

Quantum Biochemistry

ALBERTE PULLMAN AND BERNARD PULLMAN

Institut de Biologie Physico-Chimique, University of Paris (France)

1. Introduction

The purpose of quantum biochemistry is to apply the general ideas and methods of wave mechanics to the study of the electronic structure of biological molecules in relation to their behaviour as substrates of life and to their involvement in the biochemical and biophysical processes characteristic of living matter. A similar penetration of quantum-mechanical ideas and methods has already been accomplished some time ago in organic and physical organic chemistry and has resulted in an extraordinary enrichment of these disciplines, in fact in a profound transformation of what was essentially a highly developed technique into a real branch of science. The enrichment concerned in particular the deeper understanding in the *appropriate* terms of the nature of the electronic factors responsible for the structural and dynamic properties of organic compounds. It resulted in new concepts, and a number of daring suggestions and predictions which all stimulated the remarkable blooming of that discipline.

It may therefore seem strange, at first sight, that the similar penetration of these powerful and obviously so fruitful ideas and techniques into biochemistry has been delayed for a long time and, in fact, only started in a systematic and quantitative way a few years ago. The more so as many biologically highly important molecules are simply organic molecules of a medium degree of complexity.

The explanation of this puzzle lies probably, in the first place, in the mutual ignorance of biochemistry by quantum chemists and of quantum chemistry by biochemists. Following my personal opinion the fault in this respect is greater on the side of the quantum chemists than on that of the biochemists. In the second place, it is obvious that biochemistry is, in fact, infinitely more complex than organic chemistry if only because of the involvement in it, in its very central mechanisms, of huge macromolecules and because of the close interdependence of its processes. Consequently, a successful extension of quantum chemistry into biochemistry does require the existence of an already well-established quantum organic chemistry.

The inevitable, however, always ends by happening. The recent tendency of transforming biology itself into a *molecular* science, the realization that the fundamental phenomena of life are embodied in the properties of a series of fundamental molecules or macromolecules and in their interplay, have still enhanced the need for the elucidation of the electronic characteristics of these compounds and of the role of the electronic factors in their behaviour. In fact, quantum biochemistry started practically at the same time as modern molecular biology, in the 1950's. Because it came late into a field which has been awaiting its penetration the progress has been very rapid.

Besides its obvious merit of *interpreting the observed experimental facts in terms of the appropriate fundamental physical entities which correspond to these facts*, the application of quantum-mechanical methods to the study of the electronic structure of biomolecules or the investigation of biophysical and biochemical problems presents three complementary advantages.

The first advantage is the general, *universal character* of the method, its *unlimited applicability*. Thus, the usual experimental methods of physics and chemistry are intended to study essentially one (sometimes more but never too many) specific molecular property. Each such method can therefore give only a partial view of the molecular reality. The situation is quite different in the quantum-mechanical studies: in these procedures a single calculation, the solution of the wave equation, leads to a multiplicity of results which, *in principle*, (*i.e.* if we were really able to solve rigorously the extensive equation) yields the complete information about *all* the structural properties of the atomic or molecular system under investigation. In fact, even if, as it is the case *in practice*, we can only solve approximately somewhat reduced equations, the amount of information, while of course approximate and partial, still generally covers a wide variety of aspects of the problem studied.

It may reasonably be supposed that because of the complexity of biochemical systems and problems, such a multivalent approach may be particularly useful.

The second advantage of the quantum-mechanical approach to biochemistry and biophysics resides in the possibility that it offers to *precede* experimentation in a number of fields in which this experimentation seems to be particularly difficult to carry out. Thus, the calculations frequently permit determining the values (more or less exact values, according to the degree of refinement of the calculations) of a series of physicochemical characteristics of molecular systems which seem to be at present beyond the possibilities of experimental determination, or which are at least very difficult to measure presently. Among these characteristics are *e.g.*, dipole moments, ionization potentials, electron affinities, resonance energies, etc., all of which are fundamental quantities for the understanding of the physicochemical properties of molecules. The calculation of these quantities frequently permits discovery and prediction of new correlations between structure and behaviour, and sometimes completely new aspects of biochemical problems.

Finally, although it is probably the ground-state chemistry (and thus the ground-state quantum chemistry) which are essentially involved in most biological phenomena, some biochemical processes, in particular those which involve radiation effects, excited states, or electron and energy migrations, can only be properly formulated and dealt with through the use of the concepts of quantum mechanics. Their understanding is therefore entirely dependent on the application of these concepts to biochemistry.

(a) The methods and the scope of quantum biochemistry

Quantum theory provides us with two fundamental methods for the study of the electronic structure of molecules: the valence-bond method, whose simplified qualitative version is referred to frequently as the resonance theory, and the molecular-orbital method. Both represent approximate procedures for obtaining approximate solutions of the Schrödinger equation relative to molecules. This equation is the basic equation of the quantum theory and its resolution provides the electronic energy levels and the distribution of the electronic cloud in chemical systems. Approximate procedures are needed because we are unable, at present, to solve rigorously the Schrödinger equation for any atomic or molecular system beyond the very simplest ones.

Both methods have met with outstanding success in the field of organic chemistry. The predominant importance of their contribution is due, however, to different types of achievement. Thus, the resonance theory has been of great value for the general development of our ideas about the electronic structure of molecules, and for the penetration of these ideas among organic chemists, because of its availability in a non-mathematical form and its constant use of concepts with which chemists are well acquainted. Because of great mathematical complexities its quantitative utilization was, however, limited to very simple hydrocarbon systems. The molecular-orbital method, on the other hand, has been extensively applied to quantitative calculations on the electronic structure of even relatively complex organic molecules. It appeared therefore very quickly that it is this last method which presents the greatest advantages in favor of its utilization in biochemistry[1]. In fact, practically all work done till now in quantum biochemistry is essentially based on the molecular-orbital approach[2].

The limits imposed upon the dimensions and the scope of this chapter do not permit us to go into all the details of the description of the method. A general review of its directing ideas and of the branching of the different approximations and their interconnections has been given recently[3]. A general review of the principles of the molecular-orbital method has also been presented in ref. 4. The greatest part of the work carried out in applying the quantum theories to biochemistry was concerned with the so-called conjugated systems, rich in π-electrons, and involved the utilization of a semi-empirical approximation, sometimes referred to as the Hückel approximation*. This approximation has been succinctly described in volume I of this series[5] and a more extensive description of it, specially adapted for biochemists may be found in ref. 2 (pp. 3–181). Although limited originally largely to the study of the π systems of the conjugated biological molecules, the applications of the theory have more recently been extended so as to encompass the σ electrons[6]. Calculations have thus been carried out for saturated biomolecules, in particular the α-amino acids of proteins[7,8] and for the σ skeleton of the conjugated biomolecules[9], an extension which has improved the study of such properties as the dipole moments or tautomeric equilibria[10]. Representative developments have also been made in the study of the effect of non-planarity on the electronic properties of conjugated biomolecules[11], an important problem in a number of cases.

Presently let us give an outline of the main features of the method, particularly as oriented towards the study of conjugated molecules.

The basic idea which lies at the foundations of the method of molecular

---

* Although recently extensive applications have also been made of more refined approximations. See *e.g.* ref. 159.

orbitals is a very general one which has been first used in the quantum-mechanical description of polyelectronic atoms. It consists of *constructing the wave function of a polyelectronic system as a suitable combination of individual one-electron wave functions.*

The most "suitable" combination has been shown to be of the general form

$$\begin{vmatrix} a(1) & b(1) & c(1) & \ldots \\ a(2) & b(2) & c(2) & \ldots \\ \vdots & \vdots & \vdots & \\ a(n) & b(n) & c(n) & \ldots \end{vmatrix}$$

a notation which stands for the determinant built with the n individual wave functions a, b, c, etc. Since each of these is a product of an "orbital" part

$$\varphi(x, y, z)$$

and a spin function α or β, the total wave function for an even number of electrons is written as:

$$\Psi = (n!)^{-\frac{1}{2}} \begin{vmatrix} \varphi_1(1)\,\alpha(1) & \varphi_1(1)\,\beta(1) & \varphi_2(1)\,\alpha(1) & \varphi_2(1)\,\beta(1) & \ldots \\ \varphi_1(2)\,\alpha(2) & \varphi_1(2)\,\beta(2) & \varphi_2(2)\,\alpha(2) & \varphi_2(2)\,\beta(2) & \ldots \\ \vdots & \vdots & \vdots & \vdots & \\ \varphi_1(n)\,\alpha(n) & \varphi_1(n)\,\beta(n) & \varphi_2(n)\,\alpha(n) & \varphi_2(n)\,\beta(n) & \ldots \end{vmatrix}$$

Such a "Slater determinant", as it is often called, would in fact be the correct wave function for a system of non-interacting electrons. Electrons, however, do interact in real molecular systems. Thus in order to obtain a satisfactory representation, *the individual orbitals φ are determined so as to take into account the presence of the other electrons.* This is the principle of the method of molecular orbitals.

The best procedure which allows the determination of the individual molecular orbitals is the "self-consistent field" method, whose main features are as follows:

(*a*) One writes the exact *total* Hamiltonian for the system with explicit inclusion of electron interaction:

$$H = \sum_{v} H(v) + \sum_{\mu < v} \frac{1}{r_{\mu v}}$$

where $H(v)$ is the Hamiltonian for *one* electron v in the field of *all* the *bare* nuclei.

(b) One expresses the *total* energy of the system by the standard quantum-mechanical expression

$$\varepsilon = \frac{\int \Psi^* H \Psi \, d\tau}{\int \Psi^* \Psi \, d\tau}$$

in terms of the *individual* orbitals φ, by using the determinantal expression of Ψ.

(c) One satisfies the variation principle for the energy. This is a standard procedure which rests on a fundamental theorem of quantum mechanics, namely that the energy calculated by the above expression using an approximate wave function lies always higher than the exact energy; thus if one uses an approximate wave function expressed in terms of certain parameters, minimization of the energy with respect to the parameters will yield the best possible energy value attainable with this form of Ψ. Carrying out this program yields the general "Fock" equations, one for each individual orbital φ

$$F\varphi_i = \varepsilon_i \varphi_i$$

where F is an operator playing the role of an individual Hamiltonian and ε_i is the individual energy of one electron occupying the orbital φ_i.

The drawback of the Fock equations resides in the fact that each individual operator F depends on all the orbitals which are occupied in the system (on account of the explicit inclusion of the interaction terms): thus each φ is given by an equation which depends on all the φ's. The way out of this difficulty is to choose arbitrarily a starting set of φ's, calculate the $F(v)$'s, solve the series of equation for a new set of φ's and go over the same series of operations again and again until the nth set of φ's reproduces the $(n-1)$th set to a good accuracy. Hence the name "self-consistent" given to the procedure. The orbitals obtained in this fashion are in principle the best possible orbitals compatible with a determinantal Ψ.

One restriction must be made, however, about this last statement: a choice must be made of a starting set of φ's. Since it is impossible to guess, *ab initio*, the appropriate analytical form of a molecular orbital, one must rely on a "reasonable" possibility. Thus the final orbitals are the best possible orbitals *of the form chosen*.

The choice of the starting orbitals rests on the following idea: suppose that we deal with a chemical bond formed between two monovalent atoms A and B by the pairing of their valence electrons, one on A, the other on B. It is natural to assume that when one electron in the molecule is close to nucleus A, its molecular orbital will resemble the atomic orbital that it would occupy in A, and a similar situation would occur in the vicinity of B. This leads to the idea that the molecular orbital may be approximated by a linear combination:

$$\varphi = c_1 \chi_A + c_2 \chi_B$$

where the χ's are the atomic orbitals.

The idea can of course be extended to a polyatomic molecule and generalized so that each molecular orbital in a molecule is a linear combination of all the atomic orbitals occupied by the electrons in the constituent atoms:

$$\varphi = \sum_r c_r \chi_r$$

This is the LCAO approximation (linear combination of atomic orbitals) of the molecular-orbital method.

Given this convention, the Fock equations yield the unknown coefficients c_r by an iterative procedure provided that a starting set c_r^0 is chosen. The practical calculation, however, involves the tedious evaluation of a large number of integrals, a number which increases so rapidly with the number of electrons that for large molecules, complete self-consistent field calculations are out of question, and will be so, most probably, for a long time to come. Fortunately, however, great simplifications are possible for a whole class of large molecules to which so many biochemicals belong, the class of *conjugated* molecules, which are usually defined as molecules which contain double bonds separated from each other by not more than one single bond. Such molecules are well exemplified by unsaturated carbon compounds such as the cyclic aromatic hydrocarbons or polyenes.

A very special feature of the structure of such compounds is that all their atoms lie in one plane, the molecular plane, in which can be drawn all the conventional single bonds. Such an arrangement can be satisfactorily

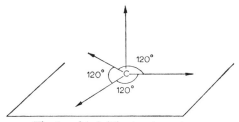

Fig. 1. sp² hybridized carbon atom.

accounted for by assuming that all the carbon atoms are in their sp^2 valence state in which the valence electrons occupy three equivalent coplanar hybrid orbitals and one p_z orbital perpendicular to the three others (Fig. 1). Under such circumstances and if maximum overlap of the electron clouds is required for the best binding, the structure of ethylene appears as shown in Fig. 2, where the second bond of the conventional double bond is seen to stem from the lateral overlap of the two p_z atomic orbitals. In an analogous fashion butadiene may be schematized by a coplanar skeleton of simple bonds, with one p_z orbital left on each carbon (Fig. 3). Here, the overlap of two adjacent p_z orbitals can not be considered as occurring on one side only, so that one can not speak of any "localized" double bonds between two definite atoms.

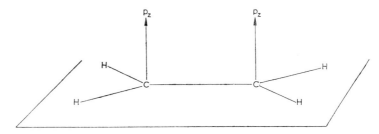

Fig. 2. The binding in ethylene.

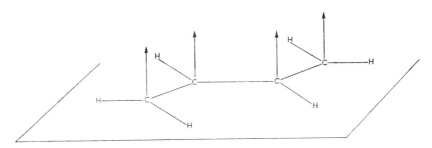

Fig. 3. Binding in butadiene.

One must speak of a "delocalized" electron system, comprising the four electrons in butadiene, and stemming from the interactions of the four atomic p_z electrons. This may be given a more quantitative aspect by saying that in the molecule each of these four electrons occupies a molecular orbital extending over the whole molecular system, and according to the LCAO concept, that this molecular orbital may be written as a linear combination of atomic p_z orbitals of the type:

$$\varphi = \sum_r c_r \chi_r$$

The remaining single bonds can be described by molecular orbitals using appropriate combinations of the atomic orbitals other than p_z. That this separation occurs stems from the peculiarity of the p_z orbital of having the molecular plane as a nodal plane. The molecular orbitals resulting from their sole combination will have the same nodal plane. They are the so-called π orbitals, the remaining molecular orbitals being the σ orbitals.

The σ–π separation which occurs in the wave functions as a result of symmetry, as well as the well-known chemical and physico-chemical evidence that the electrons of the "double bonds" are responsible for the most outstanding properties of conjugated molecules, has encouraged the practice of treating the system of π electrons alone in the field created by the nuclei *and* the so-called "σ core". In other words one writes a Hamiltonian *for the π electrons*:

$$H_\pi = \sum_v \overset{\text{core}}{H(v)} + \sum_{\mu < v} \frac{1}{r_{\mu v}}$$

defining a *core* which includes everything but the π electrons. Provided H is so defined, the self-consistent procedure can be applied to the π system alone and the best π orbitals thus computed using the Fock equation appropriate for the system. When the molecular orbitals are taken as linear combination of the atomic p_z orbitals only, the form taken by the equations amounts to solving a determinantal equation:

$$\left| F_{pq}^\pi - \varepsilon S_{pq} \right| = 0$$

with

$$S_{pq} = \int \chi_p \chi_q \, dv$$

and where the matrix elements F_{pq}^π bearing over the atomic orbitals $\chi_p \chi_q$ are defined from a Hartree–Fock operator appropriate for π electrons by

$$F^\pi = H^{\text{core}} + \sum_j (2J_j^\pi - K_j^\pi)$$

where j runs over π orbitals only, and where J_j and K_j are the Coulomb and exchange operators defined by their operation on a given function by:

$$J_j(\mu)f(\mu) = \left[\int \varphi_j^*(v)\varphi_j(v)\frac{1}{r_{\mu v}}\,dv_v\right]f(\mu)$$

$$K_j(\mu)f(\mu) = \left[\int \varphi_j^*(v)f(v)\frac{1}{r_{\mu v}}\,dv_v\right]\varphi_j(\mu)$$

Solving the equation in ε yields the possible values of the individual energies and for each ε_i the corresponding coefficients c_{ir} of the atomic orbital χ_r in the molecular orbital φ_i are obtained by a system of linear equations

$$\sum_q c_{iq}(F_{pq}^\pi - \varepsilon_i S_{pq}) = 0$$

The resolution of the equation yielding ε can be achieved by an iterative procedure of a "self-consistent" character, since all the quantities involved in F_{pq} can be, in principle, calculated when the σ core is assumed to have a definite configuration. In practice the σ core is admittedly taken as the sum of the atomic cores with the σ electrons occupying atomic valence-state orbitals.

This purely theoretical scheme yields, in principle, the best possible LCAO molecular orbitals for a given problem. However, the practice of the calculation forces in a few errors, one of which is the use of a valence-state σ core instead of an exact σ core, rarely available. Another is the choice and the limitation of the basis of atomic orbitals on which one builds the molecular orbitals. Last, but not least, there is the intrinsic error contained in the concept itself of doubly occupied orbitals which permits two electrons of opposite spins to be at the same time at the same place, thus ignoring at least in part what is called their "correlation".

In order to make up for these defects of the self-consistent field method some empirical or semi-empirical corrections must be introduced, the calibration of which are obtained by a close comparison of the theoretical results with experimental data for some fundamental compounds. The resulting procedure in its current form is the so-called Pariser–Parr–Pople approximation of the SCF method. The results obtained can still be improved by configuration mixing correcting for the residual correlation error.

The carrying out of this procedure yields the individual energies of the M.O.'s in absolute values, that is, in electronvolts. The total energy of the system is also given in absolute value. It must be remarked that the total energy does not appear here as a simple sum of individual energies since electron interaction is correctly introduced in the equations both at the individual and at the total level.

There is another approach to the determination of individual molecular orbitals, which is different in its spirit from the preceding one in that it forgets about the apparent rigor of the self-consistent formalism (apparent in so far as it must be empiricized). Thus, in this approach, instead of trying to determine the *best* molecular orbitals of an LCAO form which minimize the energy corresponding to a determinental wave function one looks for *approximate* molecular orbitals (always of an LCAO form) in the following way: one considers that each π electron of the systems moves in an "effective" field resulting from the field of the σ core including the nuclei and the averaged repulsions of the other π electrons. Defining the corresponding individual "effective Hamiltonian" H_{eff}, the solving of an *individual* Schrödinger equation:

$$H_{\mathrm{eff}}\varphi = E\varphi$$

yields the individual energy E and orbital φ.

If one looks for a molecular orbital of an LCAO form, one is led to the equations:

$$\sum_s C_s(H_{rs} - ES_{rs}) = 0$$

with the definitions:

$$H_{rs} = \int \chi_r^* H \chi_s \, d\tau$$

$$S_{rs} = \int \chi_r^* \chi_s \, d\tau$$

where H is the individual effective Hamiltonian. The individual energies are solutions of the equation:

$$|H_{rs} - ES_{rs}| = 0$$

Formally, these equations are similar to the SCF LCAO MO equations. Thus, a good definition of $H_{\mathrm{effective}}$ may yield satisfactory solutions, without the tedious iteration procedure of the SCF method. In practice, two calculation procedures have evolved from these equations, the Hückel approximation and the Wheland–Mulliken approximation. Neither of them specifies the analytical form of H, but instead treats some of the matrix elements H_{rs} as *adjustable* parameters. Their main difference resides in the neglect (Hückel) or non-neglect (Wheland–Mulliken) of overlap. Their common feature is the *tight-binding approximation*, namely the neglect of all matrix elements which involve non-bonded atoms.

The interconnections of the two approximations have been extensively discussed. From a theoretical point of view, calculations without overlap (Hückel type) can be considered as a way to reproduce with less labor the results of the calculations with overlap (Wheland–Mulliken type) provided that the quantities H_{rs} are interpreted as bearing over an orthogonalized basis[3]. Thus, in practice, the Hückel approximation is used most frequently with parameters $H_{rr} = \alpha_r$ and $H_{rs} = \beta_{rs}$ along each pair of adjacent atoms. These parameters are all expressed in terms of the corresponding values α and β for a standard carbon compound which need not be explicited until the end of the calculation. The individual energies of the electrons are finally obtained in the form

$$E_i = \alpha + k_i \beta$$

for each molecular orbital φ_i, where k_i is a number. Electrons are fed into the available orbitals by pairs with antiparallel spins as in the SCF procedure. This leaves a certain number of non-occupied orbitals in the ground state, which can serve to build excited states or negative ions of the molecule considered.

In the simple method the total Hamiltonian is admitted to be a simple sum of the individual effective Hamiltonians and thus the total energy is a simple sum of individual energies. That this is a rough approximation is well-known since if interaction is admittedly included in each individual Hamiltonian, a simple summing up will include it twice in the total energy. However, it must be recalled that since all energy quantities are expressed in terms of a certain β unit, the use of a different empirically determined unit for individual and total energies may adequately correct for this defect.

Whatever procedure is used to calculate the molecular orbitals and their energies, a certain number of electron and energy indices can be defined from them, which are related to the essential chemical, physicochemical and biochemical properties of molecules. For a detailed discussion see ref. 2. Here we simply recall the main definitions:

(1) The energy of the highest occupied molecular orbital in the ground state (homo), is a good approximation to the value of the first π molecular-ionization potential (with changed sign). This is the generalization of Koopman's theorem for atomic Hartree–Fock energies.

(2) By an extension of the previous theorem, the energy of the lowest empty molecular orbital (lemo) may be taken as a measure of the electron affinity of the molecule.

(3) The resonance energy (which should better be called the *delocalization energy*) is defined as the difference between the total π energy resulting from the calculation and the energy (calculated in the same approximation) of a hypothetical structure in which the double bonds are supposed to be completely localized.

(4) The structure indices are defined from the coefficient C_{ir} of the atomic orbitals in the molecular orbitals. Let us look at one molecular orbital:

$$\varphi_i = \sum_r C_{ir}\chi_r$$

Since φ_i is the wave function of one electron in the molecule, the quantity

$$|\varphi_i|^2\,d\tau$$

is, according to the probability interpretation of quantum mechanics, the probability of finding the electron in the volume element $d\tau$. Accordingly the summation over all space of this element is unity. This may be written:

$$\sum_{r,s} C_{ir}\,C_{is}\chi_r^*\,\chi_s\,d\tau = 1$$

or

$$\sum_r C_{ir}^2 + \sum_{s \neq r} C_{ir}\,C_{is}\,S_{rs} = 1$$

If the orbital is occupied by n_i electrons we may define the "electron population" of the orbital by the quantity:

$$\sum_r n_i\,C_{ir}^2 + \sum_{s \neq r} n_i\,C_{ir}\,C_{is}\,S_{rs} = n_i$$

Let us define around each center r the quantity:

$$n_i(C_{ir}^2 + C_{ir}\,C_{is}\,S_{rs})$$

so that its summation over all atoms r gives the orbital population. If, instead, we sum it over all the occupied orbitals we obtain the "gross atomic population" or in more familiar terms the *electronic charge* (in e units) around atom r:

$$Q_r = \sum_i n_i(C_{ir}^2 + C_{ir}\,C_{is}\,S_{rs})$$

In both the Pariser–Parr and the Hückel procedure this reduces to

References p. 56

TABLE I

PRINCIPAL APPLICATIONS OF THE ELECTRONIC INDICES

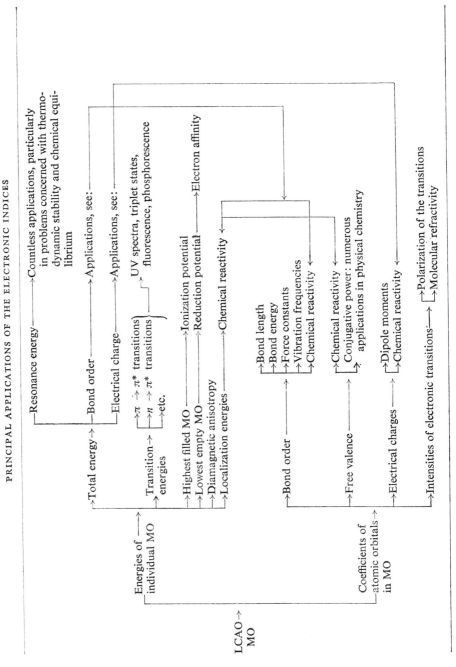

$$Q_r = \sum_i n_i C_{ir}^2$$

On the other hand the "bond order" of a bond rs may be defined as:

$$p_{rs} = \sum_i 2C_{ir} C_{is}$$

Then summing p_{rs} over all bonds departing from a given atom r gives the "bond number" characteristic of r and the difference between this bond number and the maximum possible value for all molecules gives the residual "*free valence*" on atoms r

$$F_r = N_{max} - \sum_{s \text{ adjacent to } r} p_{rs}$$

A general schematic view of the correlation of the calculated indices and the observable characteristics is given in Table I (for more details, see ref. 2). The possibilities of the methods seem, in principle, unlimited. Many practical difficulties, however, are still encountered in the calculations as we have briefly mentioned in the previous description. Thus as outlined in details elsewhere (ref. 2, pp. 178–181 and ref. 12), it is still preferable to consider that the calculations are particularly well suited for the *comparative* study of the electronic structure of molecules, in other words for the classification on a *relative* scale of compounds or molecular regions or their constituent atoms with respect to the electronic properties under investigation. We are, however, steadily approaching the stage at which the *absolute* values of the quantities evaluated, at least of some such quantities, become quite reliable.

(b) Present status of the calculations

The quantum-mechanical calculations which have so far been performed in the field of biochemistry may be considered from two viewpoints. In the first place, they may be classified with respect to the different groups of biomolecules to which they refer. Secondly, they may be classified in connection with the general biochemical problems that they have dealt with. The two classifications overlap, of course, general problems being sometimes related to a specific group of molecules and sometimes involving compounds belonging to different series. Table II presents a classification which sums up the essential quantum-mechanical calculations which have been performed so far in connection both with the electronic properties of the principal groups

TABLE II

PRINCIPAL GROUPS OF BIOMOLECULES AND PRINCIPAL GENERAL PROBLEMS IN BIOCHEMISTRY WHICH HAVE BEEN DEALT WITH, SO FAR, BY QUANTUM-MECHANICAL CALCULATIONS

Principal groups of molecules	References	Problems of particular general interest	References
Purines and pyrimidines	2, 5, 12–14, 29, 30	Hydrogen bonding	12, 16, 17, 20, 95
Purine and pyrimidine pairs of DNA	2, 10, 12, 15–17, 150	Mechanisms of mutagenesis	19–21, 58, 151
Pteridines	2, 5	Semiconductivity	12, 22–24, 26, 28, 31
α-Amino acids and proteins	7, 8, 23–28	Electron donor and acceptor properties, charge transfer	2, 12, 60–64, 92, 152
Porphyrins, iron porphyrins and hemoproteins	2, 32–37, 43	Free radicals in biology	2, 45, 65, 73, 153
Carotenoids and retinenes	2, 38	Radiation effects and the mechanism of radioresistance	2, 18, 56, 57
Energy-rich phosphates	2, 39, 40–42	Mechanisms of photobiology	2, 38, 66
Quinones	2, 44	Mechanisms of carcinogenesis	67–69
Oxidation–reduction coenzymes	2, 32, 45–49	Biochemical evolution and the origin of life	70, 71
Folic acid coenzymes	2, 50, 51	Electronic aspects of pharmacology	72–78, 154–158
Pyridoxal phosphate	2, 52, 53	Intermolecular forces	12, 29, 79–84, 93, 94, 150
Thiamine pyrophosphate	2, 54, 55	Optical properties of biopolymers	85–91
Vitamin B_{12}	59		

of biomolecules and with a number of general problems of biochemistry that they have dealt with. The list is, of course, not exhaustive.

It is not possible in this review even to outline the general results concerned with all these differents groups of biomolecules and problems. We have therefore chosen a few representative topics, the discussion of which illustrates the general scope of the approach.

2. Aspects of the electronic structure of the nucleic acids

(a) Overall results

Nucleic acids are so much at the very center of modern biochemistry and are so intimately related to a number of fundamental biological problems —heredity, protein synthesis, mutagenesis, carcinogenesis, radiation sensitivity, etc.—that it is certainly not astonishing that they have also been subjected to the quantum-mechanical approach, and that quite a number of calculations are therefore available now in connection with different aspects of their electronic structure. In fact, the nucleic acid themselves, being extremely huge aperiodic polymers, difficult to deal with explicitly, the calculations have been mostly restricted to the evaluation of the electronic characteristics of the essential components of these macromolecules, namely the purine and pyrimidine bases and the purine–pyrimidine complementary pairs, although some simplified calculations have been carried out on some electronic properties of the macromolecules themselves (e.g. semiconductivity). Particularly illustrative are the calculations carried out for the "miniature" nucleic acid formed by the two complementary base pairs, adenine–thymine and guanine–cytosine, bound together by the sugar–phosphate linkages, as illustrated in Fig. 4. In this model the interactions through the hydrogen bonds between the purine and pyrimidine of each pair are taken into account but not the interactions between the two stacked pairs. Such an approximation corresponds obviously to the viewpoint that a certain number at least of the electronic properties of the base pairs are, in the nucleic acids, relatively independent of the influence of the adjacent base pairs. This, of course, is a hypothesis which was to be verified by comparison between the theoretical results obtained with this assumption and the experimental data. It has by now been abundantly shown that:

(1) the hypothesis is satisfactory for a large number of the properties of the nucleic acids; (2) abundant information may be drawn from the calculated

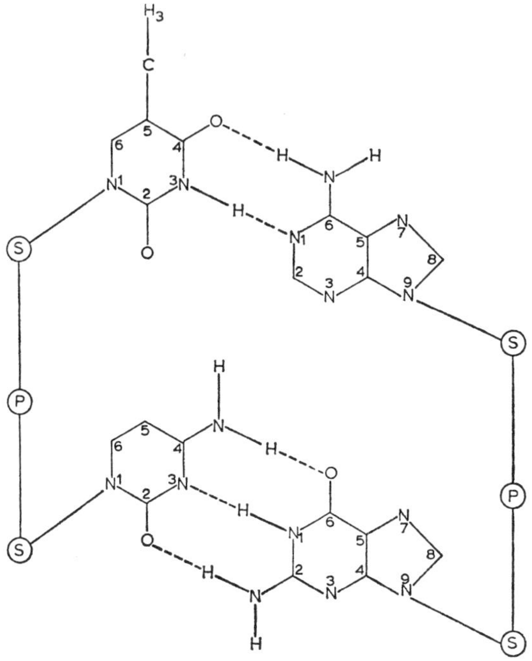

Fig. 4. "Miniature" nucleic acid.

properties of the base pairs about the essential aspects of their mutual inter-action.

It is essential to stress that our present theoretical knowledge about the electronic properties of the nucleic acids and their constituents arises from many calculations carried out by different techniques, and that although the exact values of the calculated electronic indices depend, of course, on the particular method utilized, the comparison of the different results clearly shows that they all lead to substantially the same conclusions in so far as the general aspects of the electronic structure are concerned. In particular, the calculations make possible, naturally, the determination of the sites at which the different indices have their most significant values. This is, frequently, one of the most important aspects of the results, because this determination enables one to locate the essential sites associated with the corresponding physico-chemical properties of the nucleic acids. Now, all the methods so far utilized for quantum-mechanical calculations on the nucleic bases or base pairs definitely lead to practically the same conclusions in this respect. These conclusions are summarized in Fig. 5 which represents therefore the quint-essence of the quantum-mechanical calculations for the "miniature" nucleic acid defined previously.

The precise meaning of the different qualifications indicated in this figure is the following: when a qualification is underlined by a heavy line this signi-

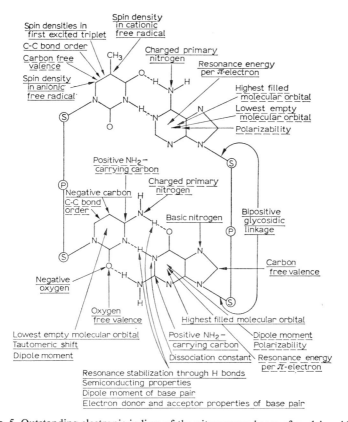

Fig. 5. Outstanding electronic indices of the nitrogenous bases of nucleic acids.

fies that it has its most outstanding value at the particular site at which it is indicated (and which, following the case may be either a base pair, or a base, or a fraction of a base, *i.e.* an atom or a bond); when the qualification is underlined by a broken line this means that this is its next most important location.

Thus *e.g.* the qualification "resonance energy per π electron" underlined with a heavy line points towards the adenine ring. This means that among the four bases of the nucleic acids and in so far as they preserve to a large extent their individuality in these acids, adenine is the one which has the greatest resonance energy per π electron. The same qualification, underlined with a broken line, is associated with guanine, which signifies that after adenine it is guanine which has the greatest resonance energy per π electron. The immediate conclusion from this situation is that the theory obviously *predicts* that the resonance stabilization of the bases, as measured by this index, should be greater for the purines than for the pyrimidines and that it should be the greatest for adenine. The reader acquainted with the concept of resonance energy may immediately guess the significance of this *theoretical prediction* for the understanding of a number of properties of the bases or even the nucleic acids, or for possible predictions in this field.

As an example of a property related to the base pairs we may consider the qualification "resonance stabilization through H bonds" which heavily underlined points to the H bonds of the guanine–cytosine pair and therefore means that the stabilization through hydrogen bonding is *predicted* to be greater for the G–C pair than for the A–T pair.

We may also consider examples of qualifications referring to the properties of more localized sites. Thus the qualification "basic nitrogen" heavily underlined is associated with N-7 of the guanine ring. This means that among all the nitrogen atoms present in the purine and pyrimidine bases of the nucleic acids, N-7 of guanine is *predicted* to be the most basic one. Again, one can easily imagine how useful such a prediction may be for the understanding of the properties of the nucleic acids which depend on basicity. Similarly, the qualification "C–C bond order" heavily underlined is associated with the C-5–C-6 bond of thymine, a situation meaning that among all the carbon–carbon bonds present in the bases of the nucleic acids this is the one which should have the highest bond order, etc.

It must be stressed that Fig. 5 is the result of purely theoretical investigations. Its real significance and usefulness depend, of course, on how far the results help in elucidating the physico-chemical and bio-chemical and -physical properties of the nucleic acids, "elucidating" meaning, of course, both the explanation of known properties and, if possible, the prediction of new ones.

A schematic representation of the principal miscellaneous applications carried out in connection with the theoretical results embodied in Fig. 5 is given in Fig. 6 which indicates the nature and the sites of the principal

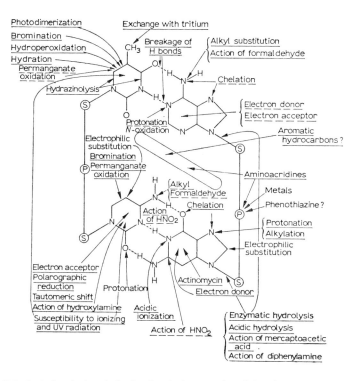

Fig. 6. Principal chemical and physicochemical properties of the nitrogenous bases of the nucleic acids.

physico-chemical properties of the nucleic acids, related to the electronic indices presented in Fig. 5. The two figures when considered together illustrate the wide and successful applicability of the theoretical evaluations, and bring into evidence the correlation between the theoretical indices and the experimental properties.

The wealth of correlations implied in these two figures precludes their detailed analysis here (for such an analysis see refs. 2 and 12). We may consider, however, a few representative examples.

(b) The significance of resonance energy

We have mentioned above the theoretical results concerned with the values of the resonance energies. In such highly conjugated heteroaromatic systems as the biological purines and pyrimidines they are extremely important

indices of electronic structure. The experimental values are however completely unknown for these molecules and their quantum-mechanical evaluation
represents thus a striking example of a theoretical calculation which is ahead
of experimental determination and is thus the basis of *predicted* correlations.
There are at least three groups of important theoretical propositions in this
field:

(*1*) The first one consists in a relationship which has been observed between
the radioresistance of conjugated compounds and the value of their resonance
energy per π electron. Not limited, of course, to the bases of the nucleic acids,
it has been shown recently to be valid for a large number of biomolecules[96].
The correlation has also been shown to be of fundamental significance for
biochemical evolution[70,71], resonance stabilization of the ground and of the
excited molecular states being possibly one of the factors orienting the
selection and conservation of biomolecules. The particular role played in
biochemistry by adenine has been linked with its particularly pronounced
thermodynamic and kinetic stability.

(*2*) The second proposition is related to the result which indicates that from
the point of view of stabilization through electronic delocalization the
guanine–cytosine pair should be more prominent than the adenine–thymine
pair. This conclusion is valid both for the relative total stability of the two
pairs and for the increment of stabilization due to the hydrogen bonding
which may be obtained by subtracting the resonance energies of the free bases
from the resonance energy of the corresponding base pairs. The differential
stabilization of the two pairs from that last point of view is equal approximately to 2.5 kcal/mole. This result when first obtained was considered as
implying that in so far as the stability of the helical configuration of the
nucleic acids may depend on hydrogen bonding, this factor should lead to a
greater stability of the acids rich in guanine–cytosine over those rich in
adenine–thymine. The greater stability of the nucleic acids rich in G–C over
those rich in A–T is verified experimentally today and is illustrated by their
higher *denaturation* or *melting temperature*[97]. It is however, also understood
today that other factors, besides hydrogen bonding, play an essential role in
this behaviour[103,104] (*vide infra*).

(*3*) Finally, the third application concerns the problem of tautomeric shifts
in the purines and the pyrimidines and is of fundamental significance in
relation to the mechanism of mutagenesis, in particular through the mis-
pairings of bases. The essential observation in this field is that in the usual
Watson–Crick pairing the bases are in their lactam and amino forms which

must thus be considered as their most stable forms. This situation is in agreement with quantum-mechanical calculations about the relative stabilities of the different possible tautomeric forms of the bases[2,10]. However, the involvement of rare tautomeric forms is considered as possibly playing an essential role in *mutations*[98,99]. The presence of a rare tautomeric form may produce an erroneous coupling, through hydrogen bonds, of unusual bases and may lead to an induced, self-perpetuating, perturbed sequence of purine–pyrimidine pairs. As the genetic information is most probably enclosed in this sequence, the result is a mutation.

Now, it can easily be shown[19] that one of the principal factors responsible for these tautomeric shifts towards rare forms and thus possibly responsible for the occurrence of mutations is the resonance energy. Thus, the essential *varying* factor responsible for the *relative* tendency of the bases to exist in a rare tautomeric form is the variation of the resonance energy which accompanies the tautomeric transformations. There are two such transformations to be considered: the lactam–lactim and the amino–imine transformations. In the lactam–lactim tautomerism, the transformation of the lactam form (the most stable) to the lactim form (less stable) is accompanied by an increase of resonance energy so that the proportion of the lactim form will be the greater, the greater this increase. In the amino–imine tautomerism, the transformation of the amino form (the more stable) to the imino form (less stable) is accompanied by a decrease in resonance energy and will therefore be the greater, the smaller this decrease. Explicit calculations carried out for the resonance energies of the different tautomeric forms of the purine and pyrimidine bases of the nucleic acids, lead then to the prediction, that the base which, from that point of view, should have the greatest tendency to exist in a rare form is *cytosine*, a prediction which seems to be confirmed from the purely chemical and physicochemical point of view (for a large list of references concerning these confirmations see refs. 10 and 19, to which may be added the more recent refs. 100 and 101). It is still more gratifying to observe that the transformation G–C→A–T seems actually to be a characteristic of spontaneous mutations[99].

(c) Van der Waals–London interactions

As already mentioned, hydrogen bonds between the base pairs are no longer considered as the major source of the stability of the double-stranded structure of the nucleic acids although they obviously play the essential role of determining the base-pairing specificity. The main source of the stability is now believed to reside in the non-specific interactions between the stacked base pairs. These are thought to consist essentially of two main components: charge-transfer forces and the Van der Waals–London interactions. We shall come back to the charge-transfer component in a later section devoted to the general problem of charge-transfer in biochemistry. Here we wish to consider only, briefly, the significance of the quantum-mechanical calculations for the elucidation of the role of the Van der Waals–London interactions. In the usual approximation of the studies of intermolecular forces, these include the electrostatic (dipole–dipole), the induction or polarization (dipole–induced dipole) and the dispersion or fluctuation (London) forces defined, in the approximation of isotropic polarizabilities by:

$$E_{\mu\mu} = \frac{1}{r^3} \left[\vec{\mu_1} \vec{\mu_2} - \frac{3}{r^2} \left(\vec{\mu_1} \vec{r} \right) \left(\vec{\mu_2} \vec{r} \right) \right]$$

$$E_{\mu\alpha} = -\frac{1}{2} \frac{1}{r^6} \left\{ \alpha_1 \left[3 \left(\frac{\vec{\mu_2} \vec{r}}{r} \right)^2 + 1 \right] + \alpha_2 \left[3 \left(\frac{\vec{\mu_1} \vec{r}}{r} \right)^2 + 1 \right] \right.$$

$$E_L = -\frac{3}{2} \frac{1}{r^6} \frac{I_1 I_2}{I_1 + I_2} \alpha_1 \alpha_2$$

The fundamental physico-chemical characteristics which enter into the evaluation of these component forces are the dipole moments, μ, the ionization potentials, I, and the polarizabilities, α, of the base pairs. The polarizabilities may be obtained relatively easily by the use of the usual additivity rules[102]. The problems of the ionization potentials and of the dipole moments are however much more difficult. As concerns the ionization potentials they are completely unknown experimentally for the biological purines and pyrimidines (as they are, in fact, for the great majority of biomolecules). As concerns the dipole moments only those of some simple derivatives of purine, adenine and uracil are known; no information exists about the moments of guanine or cytosine.

The importance of the theoretical determination of these quantities is therefore essential if we wish to understand the significance of the afore-mentioned forces for the structure of the nucleic acid and their relative contributions to the helix stability. Such determinations have been carried out by refined especially calibrated techniques for both properties and have procured very reliable values for them. Those of the ionization potentials are reproduced in Table III, both for the π and the lone-pair electrons of the bases. It can be observed that in all the bases, the lowest ionization potential corresponds to the departure of a π electron. It can also be shown that the ionization potentials of the base pairs may be taken as approximately equal to those of their purines.

TABLE III

π AND LONE-PAIR (n) MOLECULAR IONIZATION POTENTIALS IN eV

Compound	π	$n(O)$	$n(N)$
Guanine	7.6	9.4	N-7 11.2
			N-3 11.6
Adenine	8.4		N-1 11.2
			N-3 11.3
			N-7 11.4
Cytosine	8.3	9.8	11.0
Uracil	9.0	10.0 (C-6–O)	
		10.3 (C-2–O)	

Concerning the dipole moments, the striking result from the theoretical investigation is that while the dipole moments of adenine and thymine should be respectively of the order of 3.2 and 3.6 D, those of guanine and cytosine should be much greater, the predicted moments being 7.2 D for cytosine and 6.8 D for guanine. Moreover, the essential prediction is made that while the dipole moment of the adenine–thymine pair should have the small value of about 1 D, that of the guanine–cytosine pair should be many times higher, in fact correspond to about 8 D. The availability of these theoretical data makes it feasible to evaluate the contributions of the electrostatic, induced and dispersion-type attractions to the helix stability[103,104]. These contributions appear to be large. The largest terms for the interactions between the stacked bases are due to the dispersion and the dipole–dipole components. The total electrostatic, induction and dispersion free energy has been calculated for the ten different possible arrangements of the stacked complementary pairs. Its value varies between -1.6 and -19.8 kcal/2 moles of base, according to the base composition and sequence. The helix stability is calculated to be proportional to the guanine + cytosine content, which, as already mentioned, is observed experimentally.

A recent development in this field[104] makes the fundamental significance of the theoretical contribution still more important. Thus, as is well known, the usual "ideal" dipole concept only applies when the distance between the centers of the two molecules is much greater than the distance between the charge centers in both molecules. This is not the case for the stacked bases in the nucleic acids. Now, when the condition on the distances is not fulfilled, the real electrostatic forces may be quite different from what is inferred on the basis of the "ideal dipole" forces. In such a case it is recommended to forego any consideration of dipole moments and rather to think of each of the positive and negative charges in the system as interacting in a simple coulombic fashion. The electrostatic energy will then be the sum of all these monopole-monopole interactions. Now, these charges cannot be obtained otherwise, at present, than through quantum-mechanical calculations, which represent therefore the only way by which these interaction energies may be evaluated.

(d) Chemical reactivity

Before leaving this brief discussion of the possible applications of the quantum theory to the elucidation of the structure and properties of the nucleic acids a few words must be added about an aspect of this application which is in fact an example of the most typical use of quantum-mechanical calculations in biochemistry. This concerns the problem(s) of chemical reactivity. As is well known, this is a field in which quantum theory has been most successful in organic chemistry, and the same appears to be true in biochemistry. The comparison of Figs. 5 and 6 clearly indicates the extensive degree to which the molecular-orbital calculations of the electronic indices can be used for the prediction or the interpretation of the principal chemical properties of the purine and pyrimidine bases of the nucleic acids. Among the most striking successes of the theory in this field we may quote the prediction verified experimentally[105,106], that the essential center for the attack on the nucleic acids by alkylating agents should be N-7 of guanine, a prediction of significant importance for the understanding of the mutagenic activity of these agents[19]. We may also note the complete agreement between theory and experiment in fixing at the amino group of adenine the main center for the action of formaldehyde, in indicating guanine as the principal center for chelation, in indicating the C-5/C-6 bond of thymine as the site of addition reactions, in indicating the greater susceptibility of the glycosidic linkage of the purines over those of the pyrimidines towards acidic or enzymatic hydrolysis, etc.

3. Electron-donor and -acceptor properties of biomolecules and charge-transfer complexes

As already mentioned, charge-transfer complexing between the stacked bases may also make a contribution to the stability of the nucleic acids. In fact, the general problem of the occurrence of charge-transfer complexes between biomolecules and the intimately connected electron-donor and electron-acceptor properties of these compounds, are among typical problems of quantum biochemistry, the elucidation of which is to a large extent dependent on the success of the relevant calculations.

Generally speaking, charge-transfer complexes are molecular or supra-molecular entities formed from two (sometimes more) ordinarily stable molecular components through a more or less complete transfer of an electron from one of the components (the electron donor) to the other (the electron acceptor). The quantum theory of the phenomenon, which is the only satisfactory one, has been developed in the 1950's by R. S. Mulliken[107]. Following this author's proposition, the interaction of an electron-donor (D) with an electron-acceptor (A) may be described by saying that when D and A combine to form a complex, the wave function for their association may be written approximately:

$$\Psi_N = a \, \Psi_{(DA)} + b \, \Psi_{(D^+A^-)} \qquad a > b$$

for the ground state, and

$$\Psi_E = b^* \, \Psi_{(DA)} - a^* \, \Psi_{(D^+A^-)} \qquad a^* > b^*$$

for the first excited state.

In these expressions $\Psi_{(DA)}$ denotes the so called *no-bond* wave function. It means the wave function corresponding to a structure in which the binding of the two components is effected by the "classical" intermolecular forces (the electrostatic, dispersion, H-bonding, etc. forces, discussed previously), while $\Psi_{(D^+A^-)}$ denotes the so-called *dative-bond* wave function, corresponding to a structure of the complex in which one electron has been transferred from D to A and in which besides the forces listed above there may be also a weak chemical binding between the odd electrons now situated on the two components of the complex. It can be seen that the charge transfer is generally more pronounced in the excited state of the complex than in its ground state. The transition from the ground to the excited state is frequently associated with the appearance of a new absorption band, situated generally toward long wavelengths, which is the essential and practically the only unambiguous indication of the formation of a charge-transfer complex, although such a complex formation is frequently associated with the appearance of other characteristics too, *e.g.* the enhancement of semiconductivity or of some types of reactivity.

It can be shown that the transition energy corresponding to the charge-transfer band is given approximately by

$$h\nu = I_D - E_A - \varDelta$$

where I_D is the ionization potential of the electron donor, E_A the electron affinity of the electron acceptor and $\varDelta$ a stabilization term. In agreement with these considerations it is generally observed that when the formation of charge-transfer complexes involves different electron donors but the same electron acceptor, a linear correlation exists between the ionization potential of the donor and the position of the new charge-transfer band. In cases (in fact, quite numerous) in which the demonstration of the existence of a charge-transfer complex is difficult (*e.g.* when no charge-transfer band is seen), the existence of a correlation between the ionization potential of the electron donors and the degree of intermolecular association may be considered as an indication (although not as a proof) of the possible in-volvement of charge-transfer forces in the association. Sometimes, but less frequently, when the charge-transfer complexes involve different electron acceptors but the same donor, a similar linear relation exists between the stability of the complex or the position of the new band and the electron affinity of the acceptors.

During the last few years numerous authors have postulated the frequent formation of charge-transfer complexes between molecules of biochemical interest and in particular conjugated biomolecules, rich in π electrons (which are known from general quantum chemistry to have smaller ionization potentials and greater electron affinities than saturated molecules) and have postulated the involvement of such complexes both in the mechanism of biochemical reactions and in the structure of certain cellular components (nucleic acids, mitochondria, quantosomes).

Among the biomolecules which have most frequently been considered as implicated in charge-transfer complexes are:

(*1*) the essential components of the oxidation–reduction coenzymes (if not the coenzymes themselves), in particular the pyridinium (or nicotinamide) ring of the pyridine nucleotides and the isoalloxazine ring of the flavin coenzymes,

(2) the purines,

(3) the indole ring, in particular in connection with tryptophan, and, to a lesser extent,

(4) quinones, porphyrins and carotenoids.

The associations which have been most extensively studied and considered as charge-transfer complexes or at least as involving charge transfer as an important *component* in the overall bonding concern the interactions

(1) between biomolecules containing the indole ring and pyridine nucleotides or flavins,

(2) in, or between the oxidation–reduction coenzymes,

(3) between purines and a series of partners such as flavins, aromatic hydrocarbons, steroids, actinomycins, purines themselves, etc.

The detailed analysis of the data available in this field[152] indicates, however, that the definite formation of a charge-transfer complex has rarely been proved in all these examples in the sense that rarely has an absorption band characteristic of such a complex formation been really observed. An outstanding case in which such a band has definitely been seen and in which consequently we are surely dealing with a charge-transfer complexation concerns the associations formed between different electron donors and pyridinium salts as electron acceptors. In this field we must quote in particular the work of Kosower[108] on the model complexes between pyridinium compounds and iodine, on the intermediate in the dithionite reduction of NAD^+, the work of Cilento and Giusti[109] and of Alivisatos *et al.*[110,111] on the interaction of pyridinium and indole rings, and more recently the work of Shifrin[112,113] on the model compounds of the type of indolylethylnicotinamide (I) which has been extended to include the aromatic rings of the other

(I)

aromatic amino acids, and in which a correlation appears between the frequency of the charge-transfer band and the theoretical ionization potential of the aromatic electron donor. Particular attention must be paid to the recent result of Cilento and Schreier[114] who have found that while the model compounds 1-benzyl-3-carboxamide pyridinium chloride and 1-benzyl-1,4-dihydronicotinamide form an exothermic 1:1 complex which appears to be of the charge-transfer type, the association constant for the NAD–NADH system appears to be close to zero. It is the oxidized form, NAD, which is reluctant to complex and the authors attribute the situation to the fact that the pyridinium ring is probably already involved in an internal charge-transfer complex with the adenine moiety. This experiment shows the difference that may easily occur between model and real compounds. The formation of donor–acceptor complexes relating to the intramolecular association of the riboflavin and adenosine moieties of flavin–adenine dinucleotide has been demonstrated recently by McCormick and collaborators[128].

However, generally speaking, the appearance of a charge-transfer band is, as already noted, rare with biomolecules. In most cases, what has been observed are correlations between the strength or the rate of formation of the molecular association and the electron-donor or -acceptor properties (mostly electron-donor properties) of the participants, a result which may be considered as an indication of the possible involvement of charge-transfer forces, at least as an important component in these associations without, however, being a proof of it.

In view of the practically complete absence of experimental information on the ionization potentials or electron affinities of biomolecules, the corresponding data may and have been obtained from quantum-mechanical calculations, based on the molecular-orbital method. Nevertheless, calculations on the *absolute* values of these quantities similar to those which we have indicated above for the purine and pyrimidine bases of the nucleic acids, are still rare. On the other hand, there exists an extremely large body of information on the *relative* values of these quantities for nearly all the groups of important biomolecules. Such relative values are obtainable through the use of the simple Hückel approximation of the molecular-orbital method, the appropriate indices being the energy of the highest filled molecular orbital for the electron-donor capacity and the energy of the lowest empty molecular orbital for the electron-acceptor ability. The calculations yield these energies in the form $E_i = \alpha + K_i \beta$ where α is the Coulomb and β the resonance integral of the method. The values of K_i are generally in the range of 0 to 2 for the highest-filled molecular orbital and of 0 to -2 for the lowest-empty molecular orbital. The closer to zero the values of the coefficients for both orbitals the greater respectively the electron-donor or the electron-acceptor properties of the molecules. The reliability of this type of calculation for replacing the missing experimental quantities and for the relative classification of the ionization potentials and the electron affinities of biomolecules is

TABLE IV

ENERGY COEFFICIENTS OF MOLECULAR ORBITALS

Compound	Highest-filled molecular orbital	Lowest-empty molecular orbital
Purine	0.69	−0.74
Adenine	0.49	−0.87
Guanine	0.31	−1.05
Hypoxanthine	0.40	−0.88
Xanthine	0.44	−1.01
Uric acid	0.77	−1.19
Uracil	0.60	−0.96
Thymine	0.51	−0.96
Cytosine	0.60	−0.80
Barbituric acid	1.03	−1.30
Alloxan	1.03	−0.76
Phenylalanine	0.91	−0.99
Tyrosine	0.79	−1.00
Histidine	0.66	−1.16
Tryptophan	0.53	−0.86
Riboflavin	0.50	−0.34
Pteridine	0.86	−0.39
2-Amino-4-hydroxypteridine	0.49	−0.65
2,4-Diaminopteridine	0.54	−0.51
2,4-Dihydroxypteridine	0.65	−0.66
Folic acid	0.53	−0.65
Porphin	0.30	−0.24
1,3-Divinylporphin	0.29	−0.23
1-Vinyl-5-formylporphin	0.30	−0.21
α-Carotene	0.10	−0.19
β-Carotene	0.08	−0.18
Vitamin A_1	0.23	−0.31
Vitamin A_2	0.20	−0.26
Retinene	0.28	−0.26
p-Benzoquinone	1	−0.23
1,4-Naphthoquinone	1	−0.33
9,10-Anthraquinone	1	−0.44
Benzohydroquinone	0.63	−1
Naphthohydroquinone	0.41	−0.71
Anthrahydroquinone	0.23	−0.53

substantiated by correlations obtained between these two sets of data in a series of fundamental molecules for which both are known[2].

The results of the theoretical evaluation of the electron-donor and -acceptor abilities for a number of compounds representing the essential groups of conjugated biomolecules are illustrated in Table IV. They lead to the conclusion that from that point of view conjugated biomolecules may be divided into three main groups:

(1) compounds which function essentially as electron donors: purines (moderate donors with the exception of uric acid which is predicted to be a very good donor), pyrimidines (poor donors, alloxan being even predicted to act as an acceptor), α-amino acids of proteins (poor donors, with the exception of tryptophan) and reduced forms of flavins and of pyridine nucleotides.

(2) compounds which function essentially as electron acceptors: oxidized

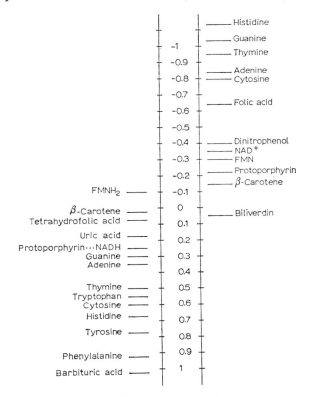

Fig. 7. Scale of electron donors and acceptors.

forms of flavins and of pyridine nucleotides, some pteridines, quinones and bile pigments.

(3) compounds which function both as electron donors and electron acceptors: porphyrins, carotenes and retinenes.

Moreover, a number of compounds of pharmacological interest, in particular the derivatives of phenothiazine, are particularly good electron donors, a situation confirmed directly by measurements of their ionization potentials[115,116].

A different presentation of these results is given in Fig. 7 which contains a *scale* of the electron-donor and -acceptor properties for some selected compounds. The electron-donor capacity of molecules is measured on the left side of the scale. The higher a compound is located on this side of the scale, the better electron donor it is. The electron-acceptor capacity is measured on the right side of the scale. The lower a compound is located on this side of the scale the better electron acceptor it is.

Very recently, these predictions concerning the electron-donor or -acceptor properties of different biomolecules have received a number of striking although sometimes indirect confirmations. This is in particular the case with the electron-donor or -acceptor properties of purines and pyrimidines. Thus, as already noted, these last types of molecule may be considered as

(II) Uric acid   (III) Barbituric acid

(IV) Alloxan

electron donors rather than electron acceptors. They appear nevertheless as moderate electron donors with the purines being on the average better donors than the pyrimidines. There are, however, exceptions to this general rule which have been *predicted* to be so[2] and they are of two kinds (Table IV).

In the first place, at least one of the purines, uric acid (II), is predicted to be an exceptionally good electron donor. On the other hand barbituric acid (III) and alloxan (IV) although playing a role in the biochemistry of the pyrimidines similar to the one played by uric acid in the biochemistry of the purines, are predicted to be particularly poor electron donors. In fact, alloxan is even *predicted* on the basis of this type of calculation to be rather an electron acceptor and a better one than the bases of the nucleic acids.

A striking although indirect confirmation of all these theoretical predictions has been brought forward recently by studies on the electrochemical behaviour of purines and pyrimidines. Thus, as far as the electron-acceptor properties are concerned, the measurements of the polarographic reduction potentials at the dropping mercury electrode[117] indicate that while purine, adenine, hypoxanthine and 2,6-diaminopurine are reducible, guanine, xanthine and uric acid are not. This is in agreement with the corresponding coefficients of the lowest-empty molecular orbital of these compounds which, in the Hückel approximation, range between -0.74 and -0.92 for the four reducible molecules and exceed -1 for the three non-reducible ones. Similarly in the series of the pyrimidines, while cytosine is very easily reducible, uracil and thymine have shown stability to polarographic reduction[118,119]. The exceptionally strong electron-accepting properties of alloxan have been entirely confirmed by similar polarographic studies on the system alloxan–alloxantin–dialuric acid[120] and through the formation of an alloxan anionic free radical by electron acceptance through charge-transfer interaction with a number of electron donors[121,122].

On the other hand, in so far as the electron-donor properties of the purines and pyrimidines are concerned, these are deducible from the results of oxidation (anodic) waves at the stationary graphite electrode. It is found[117] that simple pyrimidines *i.e.* those which, like the bases of the nucleic acids, contain only one or two amino or hydroxy substituents, do not give an oxidation wave. On the contrary, all the purines studied with the only exception of purine itself, give an oxidation wave. Moreover, the ease of oxidation increases with the number of hydroxy groups so that uric acid is the most easily oxidized of all the compounds tested. The great electron-donor ability of this last compound is also evident from its outstanding antioxidant properties[123,124]. These results are in complete agreement with our calculations which predict just this overall behaviour of the purines and pyrimidines as electron donors. The polarographic results confirm also the prediction that among the two purines of the nucleic acids guanine is oxidized easier than adenine and that, therefore, guanine is the best electron donor among the four bases of the nucleic acids. This last result is also confirmed from the studies on the charge-transfer complex formation between the nucleic acid bases, their nucleosides or nucleotides and chloranil[125,126].

Among the different applications of these considerations it may be particularly worthwhile to quote the correlations established, e.g. between the solubilizing power of purines towards aromatic hydrocarbons and their electron-donor capacity[12,69] or between the order of reactivity of free purines in solution with actinomycin and, also, their electron-donor capacity. These correlations suggest that charge transfer may represent a significant component in the forces governing these associations, which naturally involve, however, also the Van der Waals–London forces[127], π-overlap interactions[129], etc.

Another interesting application of these data concerns the base pairs of the nucleic acids. The energy coefficients calculated for the highest-filled and lowest-empty molecular orbitals in the Hückel method are equal respectively to 0.43 and -0.87 for the adenine–thymine pair and to 0.31 and -0.78 for the guanine–cytosine pair. This signifies that the guanine–cytosine pair should be at the same time a better electron donor and a better electron acceptor than the adenine–thymine pair. This result is entirely confirmed by refined self-consistent field calculations which indicate that the quantity $I_D - E_A$ should be about 1 eV smaller for the G–C pair than for the A–T pair. In so far as charge-transfer forces between stacked base pairs may contribute to the helix stability, one may expect therefore the nucleic acids rich in guanine–cytosine to exhibit greater stability than those rich in adenine–thymine. Those forces act therefore in the same direction as the Van der Waals–London ones. Calculations are in progress now in order to establish the values of their relative contribution[104].

4. The structure of proteins (and their constituents) and the problem of semiconductivity in biopolymers

Three distinct types of study have been carried out in connection with the electronic structure of proteins and their constituents. In the first place, quantum-mechanical calculations have been performed on the electronic distribution in all the twenty α-amino acids which enter into the constitution of proteins[6-8]. These calculations have been carried out for the neutral, cationic, anionic and dipolar forms of the amino acids and they represent the first and still now the most extensive example of the application of the molecular-orbital method to the investigation of the biochemical σ systems. These calculations have been utilized in particular for the interpretation of the proton shifts observed in the nuclear magnetic-resonance spectra of the α-amino acids and for the interpretation of the variation of the dissociation constants of these compounds as a function of their structure. Thus a satisfactory over-all parallelism can be observed between the proton shifts (δ_H) and the σ-electronic charges (Q_H), as well as between the shifts and the charges of the neighbouring carbons (Q_C). The results may be best presented by the general equation

$$\delta_H = -9.92\, Q_C - 133.93\, Q_H + 9.67$$

for data relating to alkaline and acidic solutions. This is in agreement with general theory, namely that the essential contribution to the proton chemical shift comes from Q_H.

The second group of studies in this field is concerned with the possible special properties of the aromatic amino acids, due to the π electrons of their conjugated residues[2]. This concerns some aspects of their chemical reactivity and, in particular, their electron-donor and -acceptor properties. An outstanding theoretical result in this last respect was the prediction (now fully substantiated by experiment[109,110,113]) that among the amino acids of proteins the best electron donor should be tryptophan.

The third, and to some extent the most interesting type of study carried out in relation to proteins is concerned with an aspect of their supramolecular structure, namely the occurrence of electronic semiconductivity owing to an overall electronic delocalization through the whole polymer resulting in the formation of energy bands.

This proposition needs some explanation. The peptide backbone of proteins is formed of a periodic repetition of small resonating units

$$-\overset{\overset{\textstyle O}{\textstyle \|}}{C}-\overset{..}{N}H-$$

which are, however, in the backbone, separated from each other by saturated carbons. Consequently no electronic delocalization can occur along such backbones. The peptide bonds are, however, also, united to each other by secondary cross linkages, namely hydrogen bonds (Fig. 8). Inasmuch as these bonds may participate in electronic delocalization and transmit conjugation effects, the existence of such a network represents the possibility of an extended electronic delocalization involving the whole molecular framework of the protein. The question is, of course, to what extent, if at all, conjugation effects may be transmitted across the hydrogen bonds. If a

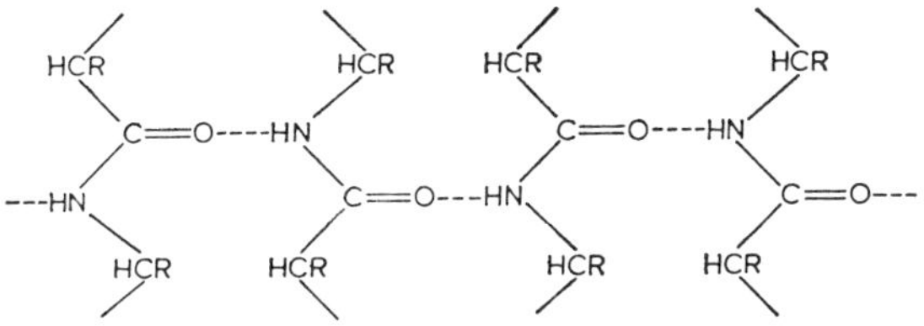

Fig. 8. The regular array of hydrogen bonds between adjacent polypeptides in a protein.

Fig. 9. The merging of molecular orbitals into bands for an infinite chain.

non-negligible transmission effectively occurs, the peptide bonds may not be considered as isolated four π-electron systems; they would fuse, more or less, following the extent of the transmission, into a giant conjugated structure. Even within the hypothesis of a weak interaction across the hydrogen bridge, quite new effects may be anticipated from such a cooperative interplay of an enormous, practically infinite, number of resonating units. It is, in particular, predictable that in these circumstances the discrete π-energy levels of the peptide will merge in the macromolecule into energy bands (Fig. 9). The problem was stated for the first time explicitly by Szent-Gyorgyi in 1941[130] and his daring hypothesis inspired a great number of theoretical (for a general review see refs. 26, 28) and experimental[131–134] researches on the semiconductive properties of proteins, and has also given rise to many discussions and even controversies.

The fundamental postulate that π-electronic delocalization can be transmitted across a hydrogen bond uniting two conjugated systems has been substantiated theoretically in recent work carried out in our laboratory[16,24,25,28], by taking explicitly into account in the calculations the empty $2p_z$ orbital on the hydrogen atom which assumes an orientation parallel to the direction of the p_z orbitals of the adjacent conjugated systems, the peptide units. Thus in a hydrogen-bonded system this concept amounts simply to extending the basis of the atomic orbitals which are linearly combined to obtain the molecular orbitals, without increasing the number of π electrons. In a bipeptide (two peptides united by an H bond) for example, the 8 π electrons of the system will occupy 4 molecular orbitals extending over the whole framework and made of the combination of 7 atomic p_z orbitals including one centered on the hydrogen nucleus (Fig. 10).

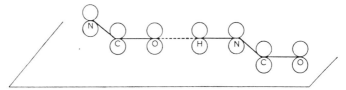

Fig. 10. Orientation of the π orbitals in hydrogen-bonded peptides.

Calculations definitely show that although each unit in a multipeptide keeps its individuality to a large extent, a certain amount of electronic delocalization does occur, with the resulting partial population of the hydrogen 2p orbital.

A number of calculations differing among themselves in technical details have been carried out on this problem. They are studied comparatively in ref. 28. A representative result obtained by self-consistent field calculations[24] is shown in Tables V and VI. Table V shows the evolution of the energies of the molecular orbitals upon multiplying the number of hydrogen-bonded peptide units, until their merging for an infinite chain into energy bands. Table VI indicates the transition energies needed for raising an electron from the filled to the empty levels (bands). The results show that four energy bands may be expected to exist in a protein chain.

The three lowest bands, including two π and one n band, are filled, and the lowest-empty is a π band. It may be noticed that the two highest-filled bands, of which one is of the π type and the other of the n type, merge partially, the upper limit of the π band being, however, nearer to the empty band than the

TABLE V

ENERGY LEVELS IN PEPTIDES (eV)

	Monopeptide π	Monopeptide n	Bipeptide π	Bipeptide n	Terpeptide π	Terpeptide n	Multipeptide π	Multipeptide n
π^*	$+1.24$	-12.63	$+1.74$ $+0.82$	-12.21	$+1.89$ $+1.32$ $+0.69$	-12.08	$+1.95$ $+0.65$	-12
π_1	-12.68		-12.02 -13.05	-13.14	-11.86 -12.38 -13.14	-12.72 -13.28	-11.08 -13.2	-13.3
π_2	-15.05		-14.55 -15.55		-14.41 -15.00 -15.63		-14.4 -15.7	

TABLE VI

ELECTRONIC TRANSITIONS IN PEPTIDES (eV)

	Triplets		Monopeptide	Singlets		
	Terpeptide	Bipeptide		Bipeptide	Terpeptide	
					9.24	
				9.22	9.22	$\pi_2 \rightarrow \pi^*$
			—9.20			
				9.21	9.22	
					7.59	
				7.63	7.55	$\pi_1 \rightarrow \pi^*$
			—7.72			
				7.54	7.45	
	7.32					
$\pi_2 \rightarrow \pi^*$	7.27	7.32				
			7.32—			
	7.25	7.25				
	5.79					
$\pi_1 \rightarrow \pi^*$	5.73	5.81				
			5.84—			
	5.69	5.73				
					5.53	
				5.42	5.44	$n \rightarrow \pi^*$
			—5.30			
				5.41	5.44	
	5.16					
$n \rightarrow \pi^*$	5.06	5.04				
			4.90—			
	5.06	5.02				

upper limit of the n band. The calculated transition energy from the highest-filled to the lowest-empty band exceeds 5 eV.

The conclusion to be drawn from these results is that inasmuch as poly-peptides can be accommodated by the proposed model, pure proteins should be rather quite good insulators. The observed semiconductivity, which, following the quoted work of Eley, Rosenberg and collaborators, involves an energy gap of 2–3 eV, is probably extrinsic, originating from extra energy levels, due to the presence of defects or impurities.

The problem of semiconductive properties is not peculiar to proteins but concerns also other types of biological macromolecules, in particular the nucleic acids. Semiconductivity, with an energy gap for dry DNA of about 2.4 eV has been observed by a number of authors[131,135,136]. It has been investigated theoretically[22] from the standpoint of the possible existence of energy bands. It may, however, also be studied from the viewpoint of a donor–acceptor mechanism of charge-carriers formation[12]. Following this general theoretical approach[137] the energy necessary for producing the separation of charges at a distance at which coulomb interactions between them are negligible is given by

$$E = I_c - E_c = I_D - E_A - P_+ - P_- = I_D - E_A - 2P$$

where I_c is the ionization energy of the crystal, E_c the electronic affinity of the crystal, I_D the ionization potential of the molecule which is the electron donor, E_A the electron affinity of the molecule which is the electron acceptor, P_+ and P_- the polarization energies due to the presence of a positive or a negative charge in the polarizable crystal. These two last energies are frequently considered as equal to the same energy P.

The quantity $I_D - E_A$ appears thus to be of fundamental importance for the determination of the energy gaps for semiconductivity, in particular for the sake of the comparison of these energies in a series of compounds, P being generally fairly constant and oscillating only feebly around the mean value of 1.6 eV.

In the absence of any experimental data in this field these quantities have been evaluated theoretically, and the application of the theory to the nucleic acids lead to a predicted value for the energy gap of the order of 4.5 eV and is greater again than the observed one. This situation suggests again that the observed semiconductivity is associated with the presence of impurities. They also lead to the prediction that the energy gap for semiconduction in DNA should decrease with increasing G–C content.

5. The origin of the "energy wealth" in the energy-rich phosphates

Why are "energy-rich" phosphates rich in energy? It means, why is the free energy of hydrolysis of certain selected phosphates a few kilocalories greater than that of the more usual ones?

A detailed theoretical investigation of the subject[41] indicates that the energy richness of the phosphates originates from a number of contributions, the most important ones being:

(1) *Opposing resonance* representing the difference between the resonance energy of the energy-rich phosphate and the sum of the resonance energies of its constituent fragments. This includes the primary opposing resonance resulting from the mere fusion of the fragments and the complementary

TABLE VII

"ENERGY RICHNESS" OF PHOSPHATES (CONTRIBUTIONS TO THEIR FREE ENERGY OF HYDROLYSIS) IN KILOCALORIES PER MOLE

Compound	Exper-imental	Funda-mental[a]	Opposing resonance		Electrostatic repulsion[b]	Keto-enol tautomerism	Free energy of ionization	Total
			primary	complementary				
ATP	7–8	3	2	0.6	+2			7.6
ADP	7–8	3	2	0.6	+1.4			7
Carboxyl phosphate	10–12	3	1.6	3	−0.7		3.2	10.1
Phosphoenol-pyruvate	11.5–12.5	3	0.2	0.8	−0.5	9		12.5
Guanidino-phosphate	9–10	3	0.4	0.8	−0.7		?	?

a Free energy of hydrolysis of the energy-poor phosphates taken as equal to about 3 kcal/mole.
b The + sign means repulsion and the − sign means attraction.

opposing resonance due to deeper structural changes produced by this fusion.

(2) *Electrostatic interaction energy.* The energy-rich phosphates are characterized by a rather unusual distribution of electrical charges, which consists
of a main backbone of at least three adjacent atoms carrying a net positive
charge surrounded by a cloud of negatively charged atoms. As an illustration,
we reproduce here the distribution of net charges in ATP (Fig. 11). This
distribution leads sometimes to strong electrostatic repulsion.

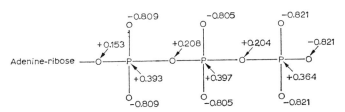

Fig. 11. The distribution of π electrons in the pyrophosphate chain of ATP.

(3) *The enol–keto tautomerism* of the product of hydrolysis.

(4) The free energy of ionization of the product of hydrolysis.

The quantitative evaluation of these contributions to the energy of the
different types of energy-rich phosphates is summed up in Table VII.

The essential contributions in each case are underlined. They are different
in each case. The general agreement with experiment is satisfactory.

6. The mechanism of enzyme and coenzyme activity

This is one field of biochemistry in which the application of the molecular-
orbital method has led to very fruitful results. The essential coenzymes are
conjugated organic molecules and the quantum-mechanical investigation of
their mode of functioning has brought into evidence the importance of
electronic delocalization in determining their ability to function as coenzymes.
The theoretical studies included the oxidation–reduction coenzymes, pyridine
nucleotides, and flavins as well as cytochromes, folic acid coenzymes,
pyridoxal phosphate, thiamine pyrophosphate, retinenes, etc. The results are
much too abundant to be even summarized here so that the interested reader
must refer to the original papers (for references see Table II). We should
simply like to emphasize the general result of this research which shows that
it is the presence of mobile electrons which, in one way or another, is
responsible for the ability of these compounds to function as coenzymes.

It has been shown, for instance, that in the respiratory coenzymes the oxidation–reduction is accompanied by an instantaneous redistribution of the energies of the molecular orbitals and, in particular of those of the lowest-empty and the highest-filled orbitals, in such a way that in each case a particularly low-lying empty orbital is associated with the oxidized form and a particularly high-lying filled orbital is associated with the reduced form. The oxidized form will then have a natural tendency to accept electrons and the reduced form to give them up, thus making these compounds particularly suitable to act as electron carriers. Such a redistribution of molecular orbitals can only occur in conjugated molecules, and, in fact, only in some very particular conjugated heterocycles.

Similarly, the studies on the group-transfer coenzymes have indicated the importance of the conjugation phenomena for the proper functioning of each of them, and have even led to the elucidation of some common features in their mechanism of action. Thus it has been shown that the essential driving force for the reactions catalyzed by these coenzymes is, practically in every case, the transformation of the primary product of the interaction of the coenzymes with a substrate into an intermediate product whose essential characteristics are, on the one hand, great energetic stability due to a high resonance energy, and, on the other hand, pronounced chemical reactivity due to the existence of large local accumulations of net, positive or negative, π-electron charges, as well as free valences. Consequently, these intermediates will display a great tendency to be formed, but also a great tendency to continue to react; in other words, they will be associated with low energies of activation for their formation but, equally, with low energies of activation for their subsequent transformations.

One or two examples will illustrate this situation. Thus following the classical work of Snell and Braunstein[138] the reactions catalyzed by pyridoxal phosphate may be schematically represented as proceeding from the initial Schiff base (V) formed between the coenzyme and an α-amino acid through transitional Schiff bases of the type VI, VII or VIII to further reaction products. It was the merit of the theory to show that the essential driving force responsible for the labilization of the groups from the α carbon of V might reside in the appreciable gain in resonance energy which is associated with the ionization. The departure of, say, the proton results in the unification of the isolated conjugated fragments of the initial base into one large resonating system. This transformation is accompanied by an increase of resonance energy of the order of 0.5 β, which represents approximately

(V)

(VI)   (VII)   (VIII)

8 kcal/mole. In fact, a similar increase in resonance energy is associated with the rupture of any of the a, b or c bonds situated around the α carbon in the initial Schiff base. It is this gain in resonance energy which may be considered as the fundamental factor responsible for the transformation of the initial Schiff base into the transitional forms and therefore for the "activation" of the α-amino acids in pyridoxal-phosphate catalyzed reactions.

Moreover, the examination of the electronic distribution in the transitional Schiff base resulting from the departure of, say, the proton yields information which is exceedingly useful for the interpretation of the further steps of the reaction. This distribution is represented in Fig. 12.

The most striking feature of this figure is the unusual charge distribution associated with the extracyclic C_{formyl}–N_{amino}–C_{α} chain. All these three atoms carry an excess of π-electron charges, so that we have a chain of three adjacent atoms all bearing net negative charges. Moreover, in spite of the fact that the greater intrinsic electronegativity of the nitrogen atom has been accounted for *ab initio* by the use of suitable parameters, the net negative charges on the formyl or α carbons are greater than the net negative charge of the amino

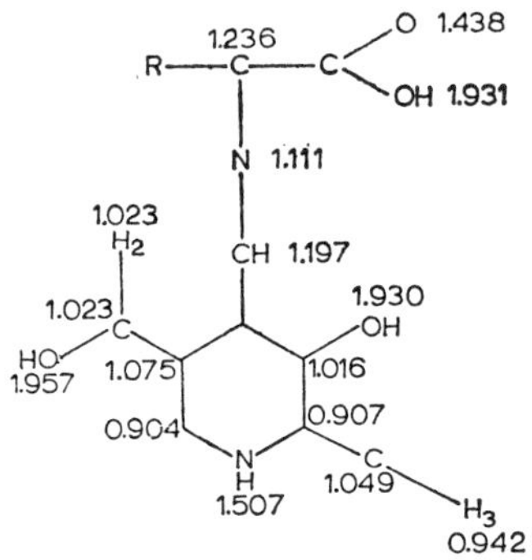

Fig. 12. Distribution of π electrons in the transitional Schiff base (VI).

nitrogen. These two carbons appear thus as the essential centers for possible electrophilic attacks on the molecule and therefore as the essential centers for future protonations. Because of their high net negative charges they are also very reactive centers. These are the very reactions which occur in VI and lead to the products of the catalysis. The protonation of VI at the α carbon leads to racemization, the protonation at the formyl carbon to transamination. A mechanism operating along the same line has been shown to govern all the reactions catalyzed by pyridoxal phosphate.

As a second example of a similar situation we may consider thiamine pyrophosphate (IX). Let us consider a typical reaction catalyzed by this coenzyme. Such a reaction (benzoin condensation) is shown in Fig. 13.

Following a theory developed by Breslow and which has recently received important experimental confirmations, the mechanism of action of thiamine involves the ionization of the thiazolium ring, through the loss of a proton from C-2 with the consecutive formation of a zwitterion:

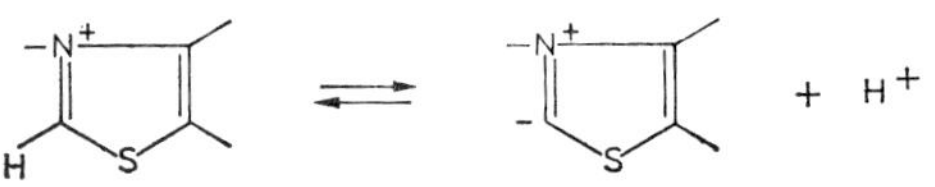

It is this ionized C-2 position which is the reactive site of the coenzyme toward interactions with the molecules to be transformed. In our example

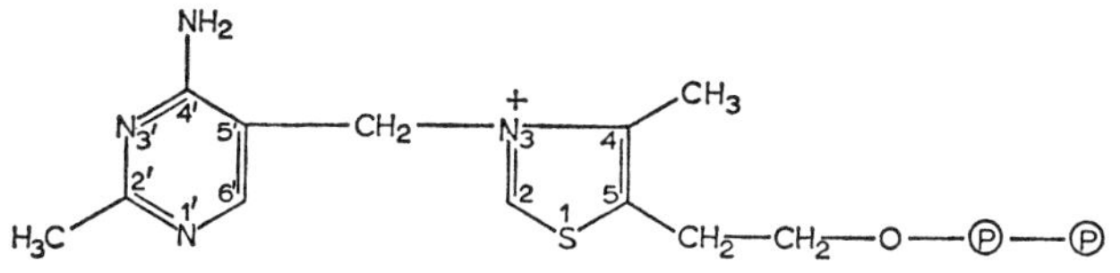

(IX) Thiamine pyrophosphate

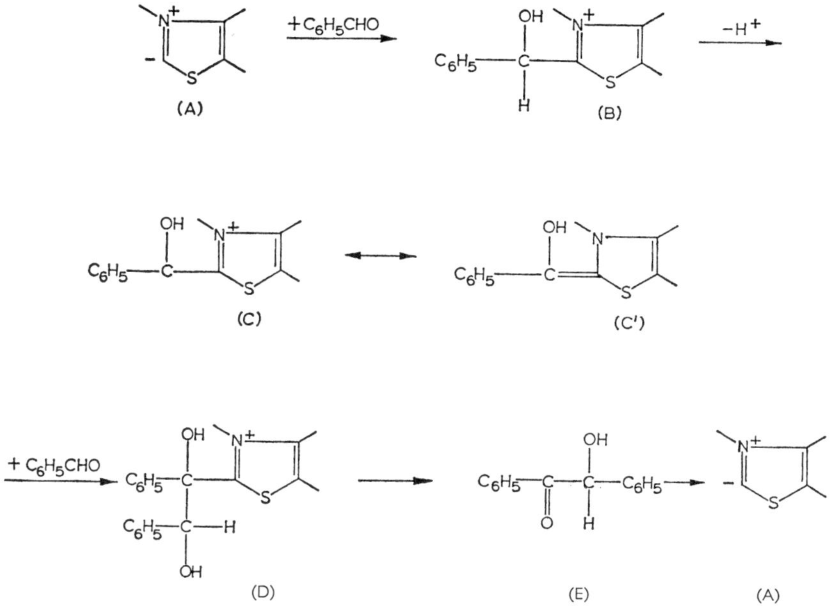

Fig. 13. The mechanism of thiamine-catalyzed benzoin condensation.

(interaction with benzaldehyde) this interaction yields the primary compound B of Fig. 13. The next step in the reaction is the release of a proton from the extracyclic aldehyde carbon with formation of the transitional but essential intermediate C, the so-called "active aldehyde". An important factor contributing towards this transformation must be the stabilization by resonance of this intermediate (which may also be represented by the other important canonical structure C'). The important point is that the release of a proton creates a unique conjugated system extending over the whole of the molecular periphery. The gain in resonance energy corresponding to this ionization is about 0.2β, *i.e.* approximately 4 kcal/mole. Moreover, as indicated in Fig. 14, the "active benzaldehyde" then possesses an extremely reactive atom in its aldehyde carbon. This carbon carries a very great excess of electronic charge: its net negative charge is 0.335e. Consequently, it will combine very easily with an electron-deficient center, such as the aldehyde carbon atom of a second C_6H_5CHO molecule (compound D of Fig. 13). The splitting of this adduct, which leads to benzoin and regenerates the thiazolium salt, is again accompanied by a gain in resonance energy, due to the establishment of resonance between the phenyl and the C=O group of benzoin and to the increase in resonance energy in the thiazole ring.

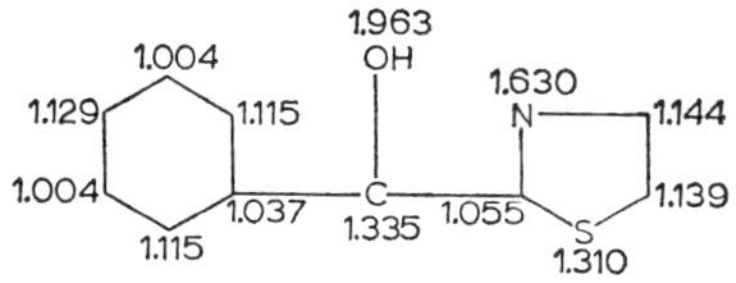

Fig. 14. Electronic charges in "active benzaldehyde".

Quite similar considerations apply to the other types of thiamine-catalyzed reactions. In particular, the structure of "active acetaldehyde" is similar to that of "active benzaldehyde".

To some extent a similar mechanism may be found in the mode of action of folic acid-dependent enzymes. The problem is somewhat more complex, and cannot be discussed in detail here. We may, nevertheless, indicate that the mechanism of transfer of one-carbon metabolic units by the coenzyme tetrahydrofolic acid (X), involves again the transformation of the initial fixation product (XI), resulting from the acceptance of the one-carbon unit,

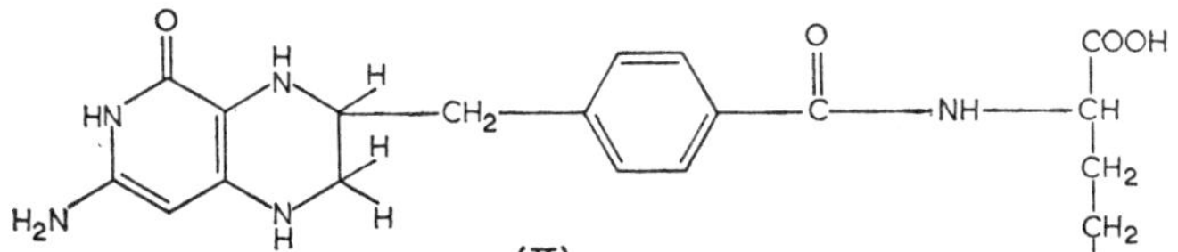

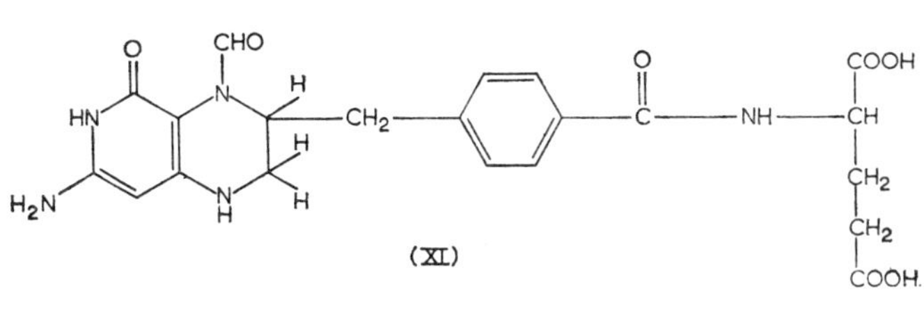

(XII)

into a more stable transitional form (XII), which is at the same time better adapted to be the one-carbon unit donor.

In conclusion, it may thus be said that the quantum-mechanical analysis carried out for the mechanism of action of the principal group transfer coenzymes enables us to establish the existence of common electronic features responsible for the catalytic activity of a number of them. These features are related to the resonance stabilization and the electronic activation of the transitional intermediates of the catalyzed reactions. This situation is in no way accidental but arises from the fact that such molecules appear to be particularly well suited to play the role of reaction sites, as it is in this type of compound that the simultaneous effects of energy stabilization and electronic activation of the transition forms may be obtained with particular ease. Their choice by nature is thus one of the manifestations of quantum effects in bio-chemical evolution.

7. The mechanism of chemical carcinogenesis

Finally, we can hardly close this review without referring to the work carried out on the correlations between the electronic structure and the carcinogenic activity of molecules. Besides its proper value, the work illustrates the possibilities of utilizing the molecular-orbital method for the understanding of the mode of action of drugs, whether harmful or chemotherapeutic.

The theoretical studies on the electronic aspects of carcinogen-esis[67-69,139,140] have been centered around two problems: (1) the correlation between structure and activity, and (2) the implication of this correlation for the interactions between the carcinogens and the possible cellular receptor sites of their activity.

(a) Structure–activity correlations in aromatic hydrocarbons

The correlations between structure and activity have been studied with different degrees of success for the different groups of carcinogens. The most successful results have been obtained for the aromatic hydrocarbons. The interest in the correlations obtained in this case is increased by the fact that aromatic hydrocarbons are most probably carcinogens *per se*, no metabolite of them being known to have activity of the same order of magnitude as the parent hydrocarbon. The correlations observed between structure and activity for these molecules may therefore be of direct significance.

One of the principal characteristics of these correlations is that although explicitly considering that the intact, completely aromatic ring system of the hydrocarbons is necessary for the appearance of carcinogenic activity, they nevertheless link the existence of this activity with the electronic properties of specific regions of the molecules. These are regions which both quantum-mechanical calculations and experimental observations indicate as being of particular importance for certain types of chemical reactivity of these compounds. They are the so-called K and L regions, whose location on the molecular periphery is illustrated in the typical example of 1,2-benzanthra-cene in Fig. 15. The theory states that in order to be carcinogenic the molecule

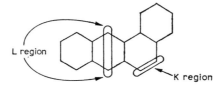

Fig. 15. Important regions in carcinogenesis.

must possess a "reactive" K region (a reactive aromatic bond) but be devoid of a "reactive" L region (reactive *para* positions). These reactivities are expressed quantitatively in terms of the "localization energies" which are the essential varying parts of the activation energies for the chemical reactions expected to occur in these two regions. The quantitative correlation which has been established is able to account, with very few exceptions, for the activity or inactivity of all the polybenzenoid hydrocarbons which have been tested experimentally and, also, to account for the relative potency of the active molecules.

Thus, the evaluation of the threshold values for the "localization energies" of the K and L regions has led to the following quantitatives rules: in order to be carcinogenic a hydrocarbon must possess a K region whose appropriate index is smaller than 3.31 β (β being the usual exchange integral of the

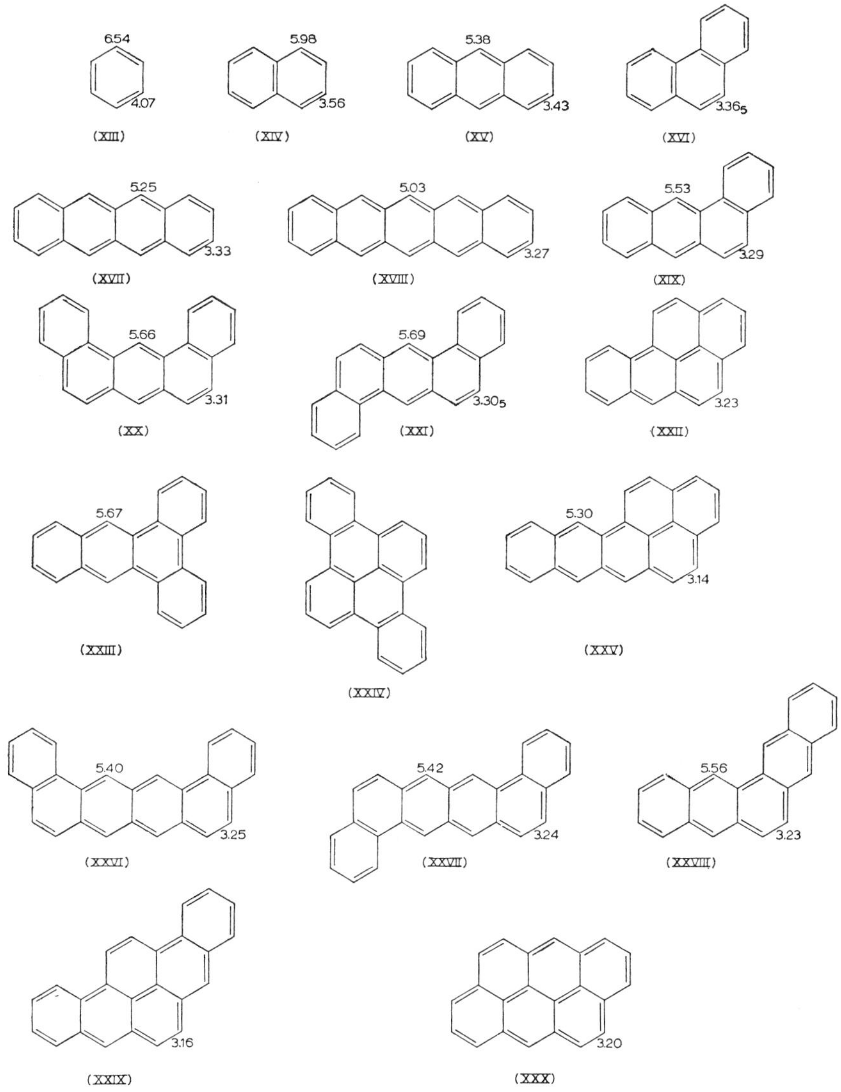

Fig. 16. The reactivity indices of the K and L regions in aromatic hydrocarbons.

molecular-orbital method); if, however, the molecule possesses also an L region it cannot be carcinogenic unless the appropriate index of that region is greater than 5.66 β. With these rules in mind let us look at a few particularly representative examples which illustrate the nature of the results obtained

(Fig. 16). Thus, none of the small molecules, benzene (XIII), naphthalene (XIV), anthracene (XV), or even phenanthrene (XVI), possess a reactive bond (a K region) whose index would be smaller than the threshold value. None of these molecules is carcinogenic. One, of them, anthracene, has a second reason for not being carcinogenic: the index of its L region is too small *i.e.* this region is too reactive.

Let us note then what happens to the indices of these two essential regions when a new benzene ring is added to anthracene linearly to give naphthacene (XV),: both indices decrease, which means that the reactivities of both regions increase. The phenomenon continues, in particular as far as the increase of the reactivity of the L region is concerned, with the linear addition of more benzene rings (*e.g.* pentacene, XVII). This result immediately accounts for one of the fundamental aspects of carcinogenesis by aromatic hydrocarbons, namely that acenes are never carcinogenic. Even if their K region is appropriate for carcinogenicity they will not be active because of the too great reactivity of their L region.

On the other hand, when a new benzene ring is added to anthracene, not in the linear but in the lateral position, to yield 1,2-benzanthracene (XIX), it may be observed that this results in a marked decrease in the index of the K region and, simultaneously, a marked increase in the index of the L region. It activates the K region but deactivates the L region and acts thus on both regions in the direction of facilitating the appearance of carcinogenicity. 1,2-Benzanthracene is considered as active by some authors but as inactive by a number of others. Its borderline position may be considered as being due to the fact that its L region is still slightly too reactive in order to be frankly compatible with carcinogenicity. When a second benzene ring is added in the lateral position to 1,2-benzanthracene as in 1,2,7,8- or 1,2,5,6-dibenzanthracenes (XX and XXI), the deactivation of the L region is increased, so that these two compounds obey the two fundamental rules which we have proposed and, thus, are both carcinogenic. The case of 3,4-benzpyrene (XXII), is particularly striking. When this compound is compared to benzanthracene, it may be seen that the inclusion of the supplementary ring increases greatly the reactivity of the K region and, at the same time, greatly suppresses the L region. The molecule is thus in a special situation for exerting a strong carcinogenic effect, which it really does.

With these examples in mind it is easier to understand the reasons for the activity or inactivity of the remaining aromatic hydrocarbons. Thus, *e.g.* 1,2,3,4-dibenzanthracene (XXIII) or 1,2,6,7-dibenzpyrene (XXIV) are not carcinogenic because their K region is blocked. 3,4-Naphthopyrene (XXV), 1,2,7,8- and 1,2,9,10-dibenznaphthacenes (XXVI and XXVII) are not carcinogenic because the addition of a supplementary benzene ring to the skeleton of 3,4-benzpyrene or of the corresponding dibenzanthracenes, creates in these molecules too reactive L regions. Pentaphene (XXVIII), although containing a K region favoring carcinogenesis, is inactive because it contains two too reactive L regions. It was predicted by the theory that the blocking of these regions, as in 3,4,9,10-dibenzpyrene (XXIX), should lead to a very potent carcinogen, a prediction confirmed by experiment. The theory accounts satisfactorily for practically all molecules of this category. In fact, the only genuine exception to the theory is anthanthrene (XXX), which although possessing a reactive enough K region and devoid of an L region is nevertheless inactive. A possible explanation may be that the two reactive "meso" carbons of the two central rings constitute a kind of an elongated, pseudo L region. Support for such a viewpoint arises from the fact that when these two carbons are blocked by methyl substituents the molecule becomes active.

The preceding correlations imply definite conclusions as to the nature of the interactions which may occur between the carcinogens and the cellular receptors and lead to the development of tumors: they suggest that the interaction leading to carcinogenesis must occur through the K region of the molecule, should most probably involve a strong chemical binding of the type of an addition reaction, possibly subject to steric hindrance when substituents are present in this region or in its vicinity. These conclusions arise from the very nature of the electronic index of the K region which is correlated with carcinogenicity. The activating effect of methyl substituents and the deactivating effect of heterocyclic nitrogens suggest the electrophilic nature of the receptor. Complementary developments of the theory, linked with the interpretation of the metabolic transformations of the hydrocarbons, suggest that the binding with the receptor should involve a quinoid type of link. On the other hand it is obvious from our fundamental propositions that interactions occurring through the L region should be unfavorable for the appearance of carcinogenicity.

(b) *Interactions between the carcinogens and possible cellular receptors*

From obvious theoretical considerations and from the available experimental data the most important cellular receptors for the action of carcinogens appear to be the proteins and the nucleic acids. The most significant results in connection with the preceding theoretical propositions are obtained, in so far as the aromatic hydrocarbons are concerned, from the study of their interactions with proteins. This study which was initiated by the Millers[141,142] and explored extensively in particular by Heidelberger and collaborators[143-148] yields, in fact, results in striking agreement with the inferences which may be drawn from the theoretical correlations. Thus they indicate the formation of a strong chemical bond, possibly of a quinoid type, between the carcinogens and the proteins. The formation of this addition complex correlates with the existence of carcinogenic activity (as indicated *e.g.* by the selectivity of its formation in the species or tissues in which the tumors develop, the inhibition of its formation by factors inhibiting tumor formation, etc.) and correlates also with its potency. In the particularly well studied case of 1,2,5,6-dibenzanthracene the reaction has been shown to occur, at least to a large extent, through the K region of the carcinogen. Studies carried out with a series of appropriately chosen compounds have shown that the interaction involved is of the type of an addition reaction, subject to steric hindrance, in the case of substituents present at the K region. A few exceptions consisting of bound non-carcinogens may probably be explained by binding through the L region or even through other centers not implicated in the theory. A very significant observation is that the binding of the carcinogens seems also to involve preferentially a specific electrophoretic fraction of the protein (which moreover is absent from the tumors), while the exceptional binding of non-carcinogens involves a different fraction. Whatever the detailed conclusions which we may be allowed to infer from this situation, it is obvious that a high degree of correlation exists between the electronic propositions of the K–L region theory, the binding of aromatic hydrocarbons to specific fractions of proteins, and the carcinogenicity of these molecules.

The interactions with the nucleic acids have been less explored, and the situation in this field is at present rather confused. Two types of interactions seem to have been observed between the aromatic hydrocarbons and these macromolecules: a weak "physical" binding and a strong "chemical" binding. The discovery of the strong "chemical" binding is very recent[149] and although it may, of course, be of fundamental importance for the problem of carcinogenesis, not enough is known about it to permit any definite or even tentative conclusion to be drawn in this respect at present. As to the weak "physical" binding rather extensive discussions are taking place presently with the forward view of determining its nature. It was shown[69], however, that whatever may be the conclusion in this respect, one fact is certain, namely, that these interactions are completely non-specific with respect to the carcinogens and are thus probably far remote from any relation with the induction of cancer.

8. Conclusions

It is difficult, if not impossible, to summarize in one chapter the principles and the essential achievements of a whole science, which although relatively young, moves forward quickly and penetrates into practically all branches of biochemistry and biophysics. The few examples of the results reached in a number of selected fields illustrate nevertheless that quantum biochemistry does fulfill the double goal which may be expected from its premises: (*1*) the elucidation of the structure and function of biomolecules in terms of appropriate fundamental physical entities, and (*2*) the guidance of experimentation by predicting the values of physicochemical characteristics of molecular systems and proposing new correlations between these characteristics and molecular behaviour. While the detailed evaluation of the electronic properties of an actual biochemical macromolecule is still a distant goal, reliable information may be deduced from studies upon simple models. The recent extension of theoretical investigations into the field of molecular interactions and associations opens the possibility for an important contribution to the problem of the mechanism of such fundamental biological reactions as duplication, coding, protein synthesis, etc.

REFERENCES

1   A. PULLMAN AND B. PULLMAN, in M. KASHA AND B. PULLMAN (Eds.), *Horizons in Biochemistry*, Academic Press, New York, 1962, p. 553.

2   B. PULLMAN AND A. PULLMAN, *Quantum Biochemistry*, Wiley–Interscience, New York, 1963.

3   A. PULLMAN, in B. PULLMAN AND M. WEISSBLUTH (Eds.), *Molecular Biophysics*, Academic Press, New York, 1965, p. 81.

4   J. I. FERNANDEZ-ALONSO, in J. DUCHESNE (Ed.), *The Structure and Properties of Biomolecules and Biological Systems*, Interscience, New York, 1964, p. 3.

5   J. H. JAFFÉ, in M. FLORKIN AND E. H. STOTZ (Eds.), *Comprehensive Biochemistry*, Vol. 1, Elsevier, Amsterdam, 1962, p. 34.

6   G. DEL RE, in B. PULLMAN (Ed.), *Electronic Aspects of Biochemistry*, Academic Press, New York, 1964, p. 221.

7   G. DEL RE, B. PULLMAN AND T. YONEZAWA, *Biochim. Biophys. Acta*, 75 (1963) 153.

8   T. YONEZAWA, G. DEL RE AND B. PULLMAN, *Bull. Chem. Soc. Japan*, 37 (1964) 985.

9   H. BERTHOD AND A. PULLMAN, *J. Chim. Phys.*, 62 (1965) 942.

10 H. BERTHOD AND A. PULLMAN, *Biopolymers*, 2 (1964) 483.

11 J. P. MALRIEU AND B. PULLMAN, *Theoret. Chim. Acta*, 2 (1964) 293, 302.

12 B. PULLMAN, in B. PULLMAN AND M. WEISSBLUTH (Eds.), *Molecular Biophysics*, Academic Press, New York, 1965, p. 117.

13 A. PULLMAN AND B. PULLMAN, *Bull. Soc. Chim. France*, (1958) 766; (1959) 591.

14 A. VEILLARD AND B. PULLMAN, *J. Theoret. Biol.*, 7 (1963) 1.

15 B. PULLMAN AND A. PULLMAN, *Biochim. Biophys. Acta*, 36 (1959) 343; *J. Chim. Phys.*, 58 (1961) 904.

16 A. PULLMAN, *Compt. Rend.*, 256 (1963) 5435.

17 P. O. LÖWDIN, *Biopolymers*, 1 (1964) 161, 293.

18 B. PULLMAN, in B. PULLMAN (Ed.), *Electronic Aspects of Biochemistry*, Academic Press, New York, 1964, p. 131.

19 A. PULLMAN, in B. PULLMAN (Ed.), *Electronic Aspects of Biochemistry*, Academic Press, New York, 1964, p. 135.

20 P. O. LÖWDIN, in B. PULLMAN (Ed.), *Electronic Aspects of Biochemistry*, Academic Press, New York, 1964, p. 167.

21 CH. NAGATA, A. IMAMURA, H. SALTO AND K. FUKUI, *Gann*, 54 (1963) 109.

22 J. LADIK, in B. PULLMAN (Ed.), *Electronic Aspects of Biochemistry*, Academic Press, New York, 1964, p. 203.

23 M. G. EVANS AND I. GERGELY, *Biochim. Biophys. Acta*, 3 (1949) 188.

24 M. SUARD, G. BERTHIER AND B. PULLMAN, *Biochim. Biophys. Acta*, 52 (1961) 254.

25 M. SUARD, *Biochim. Biophys. Acta*, 59 (1962) 227; 64 (1962) 400; *J. Chim. Phys.*, 61 (1964) 79, 89.

26 A. PULLMAN, *Biopolymers Symp.*, 1 (1964) 29.

27 S. YOMOSA, *Biopolymers Symp.*, 1 (1964) 1.

28 A. PULLMAN, in O. SINANOGLU (Ed.), *Modern Quantum Chemistry, Istanbul Lectures*, *August 1964*, Vol. III, Academic Press, New York, 1965, p. 283.

29 H. DEVOE AND I. TINOCO JR., *J. Mol. Biol.*, 4 (1962) 518.

30 R. K. NESBET, *Biopolymers Symp.*, 1 (1964) 129.

31 A. PULLMAN AND B. PULLMAN, *Biochim. Biophys. Acta*, 54 (1961) 384.

32 B. PULLMAN, C. SPANJAARD AND G. BERTHIER, *Proc. Natl. Acad. Sci. (U.S.)*, 46 (1960) 1011.

33 A. VEILLARD AND B. PULLMAN, *J. Theoret. Biol.*, 8 (1965) 317.

34 K. OHNO, Y. TANABE AND F. SASAKI, *Theoret. Chim. Acta*, 1 (1963) 378.

35 M. KOTANI, *Rev. Mod. Phys.*, 35 (1963) 717.
36 J. S. GRIFFITH, *Biopolymers Symp.*, 1 (1964) 35.
37 P. DAY, G. SCREGG AND R. J. P. WILLIAMS, *Biopolymers Symp.*, 1 (1964) 271.
38 A. PULLMAN AND B. PULLMAN, *Proc. Natl. Acad. Sci. (U.S.)*, 47 (1961) 7.
39 B. GRABE, *Biochim. Biophys. Acta*, 30 (1958) 560.
40 B. GRABE, *Arkiv Fysik*, 15 (1959) 207.
41 B. PULLMAN AND A. PULLMAN, *Radiation Res., Suppl.* 2 (1960) 160.
42 K. FUKUI, A. IMAMURA AND CH. NAGATA, *Bull. Chem. Soc. Japan*, 36 (1963) 1450.
43 M. KOTANI, in J. DUCHESNE (Ed.), *The Structure and Properties of Biomolecules and Biological Systems*, Interscience, New York, 1964, p. 159.
44 A. PULLMAN, *Tetrahedron*, 19, suppl. 2 (1963) 441.
45 B. PULLMAN AND A. PULLMAN, *Proc. Natl. Acad. Sci. (U.S.)*, 45 (1959) 136.
46 G. KARREMAN, *Bull. Math. Biophys.*, 23 (1961) 55, 135.
47 G. KARREMAN, *Ann. N. Y. Acad. Sci.*, 96 (1962) 1029.
48 B. GRABE, *Biopolymers Symp.*, 1 (1964) 283.
49 J. P. MALRIEU AND B. PULLMAN, *Theoret. Chim. Acta*, 2 (1964) 302.
50 A. M. PERAULT AND B. PULLMAN, *Biochim. Biophys. Acta*, 44 (1960) 251; 52 (1961) 266.
51 R. COLLIN AND B. PULLMAN, *Biochim. Biophys. Acta*, 89 (1964) 232.
52 A. M. PERAULT, B. PULLMAN AND C. VALDEMORO, *Biochim. Biophys. Acta*, 46 (1961) 555.
53 B. PULLMAN, in E. E. SNELL, P. M. FASELLA, A. E. BRAUNSTEIN AND A. ROSSI-FANELLI (Eds.), *Chemical and Biological Aspects of Pyridoxal Catalysis*. Pergamon, London, 1963, p. 103.
54 B. PULLMAN AND C. SPANJAARD, *Biochim. Biophys. Acta*, 46 (1961) 576.
55 R. COLLIN AND B. PULLMAN, *Arch. Biochem. Biophys.*, 108 (1964) 535.
56 B. PULLMAN AND A. PULLMAN, in M. BURTON, J. S. KIRBY-SMITH AND J. L. MAGEE (Eds.), *Comparative Effects of Radiation*, Wiley, New York, 1960, p. 105.
57 B. PULLMAN, *Mem. Acad. Roy. Belg.*, 33 (1961) 174.
58 A. PULLMAN, *Biochim. Biophys. Acta*, 87 (1964) 365.
59 A. VEILLARD AND B. PULLMAN, *J. Theoret. Biol.*, 8 (1965) 307.
60 B. PULLMAN AND A. PULLMAN, *Proc. Natl. Acad. Sci. (U.S.)*, 44 (1956) 1197; *Rev. Modern Phys.*, 32 (1960) 428.
61 B. PULLMAN, in Abstracts of the *6th International Congress of Biochemistry*, Vol. 10, New York, 1964, p. 737.
62 M. ROSSI AND A. PULLMAN, *Biochim. Biophys. Acta*, 88 (1964) 211.
63 B. PULLMAN, *Biochim. Biophys. Acta*, 88 (1964) 440.
64 A. SZENT-GYORGYI, *Introduction to a Submolecular Biology*, Academic Press, New York, 1960.
65 A. PULLMAN, *J. Chim. Phys.*, 61 (1964) 1666.
66 M. J. MANTIONE AND B. PULLMAN, *Biochim. Biophys. Acta*, 91 (1964) 387.
67 A. PULLMAN, *Biopolymers Symp.*, 1 (1964) 47.
68 B. PULLMAN, *Biopolymers Symp.*, 1 (1964) 141.
69 B. PULLMAN, *J. Cellular Comp. Physiol.*, 64 ,Suppl. 1 (1964) 91.
70 B. PULLMAN AND A. PULLMAN, *Nature*, 196 (1962) 1137.
71 B. PULLMAN, in P. O. LÖWDIN AND B. PULLMAN (Eds.), *Molecular Orbitals in Chemistry, Physics and Biology*, Academic Press, New York, 1964, p. 547.
72 B. PULLMAN, in B. PULLMAN (Ed.), *Electronic Aspects of Biochemistry*, Academic Press, New York, 1964, p. 559.
73 J. P. MALRIEU AND B. PULLMAN, *Theoret. Chim. Acta*, 2 (1964) 293.

74 G. Karreman, *Data Acquisition and Processing in Biology and Medicine*, Pergamon, London, 1962, p. 51.
75 G. Karreman, I. Isenberg and A. Szent-Gyorgyi, *Science*, 130 (1959) 1191.
76 K. Fukui, Ch. Nagata and T. Yonezawa, *J. Am. Chem. Soc.*, 80 (1958) 2267.
77 K. Fukui, Ch. Nagata and A. Amamura, *Science*, 132 (1960) 87.
78 A. Inouye and Y. Shinagawa, *Bull. Chem. Soc. Japan*, 35 (1962) 701.
79 J. Hirschfelder in B. Pullman and M. Weissbluth (Eds.), *Molecular Biophysics*, Academic Press, New York, 1965, p. 325.
80 O. Sinanoglu, S. Abdulnur and N. R. Kestner, in B. Pullman (Ed.), *Electronic Aspects of Biochemistry*, Academic Press, New York, 1964, p. 301.
81 O. Sinanoglu and S. Abdulnur, *J. Photochem. Photobiol.*, 3 (1964) 333.
82 H. Jehle, *Proc. Natl. Acad. Sci. (U.S.)*, 50 (1963) 516.
83 H. Jehle, W. C. Parke and A. Salyers, in B. Pullman (Ed.), *Electronic Aspects of Biochemistry*, Academic Press, New York, 1964, p. 313.
84 H. Jehle, W. C. Parke, R. M. Shiven and D. K. Aein, *Biopolymers Symp.*, 1 (1964) 209.
85 W. Rhodes, *J. Am. Chem. Soc.*, 83 (1961) 3609.
86 W. Rhodes, *J. Chem. Phys.*, 37 (1962) 2433.
87 I. Tinoco, *J. Am. Chem. Soc.*, 82 (1960) 4785; 84 (1961) 5047.
88 H. DeVoe and I. Tinoco, *J. Mol. Biol.*, 4 (1962) 518.
89 R. K. Nesbet, *Mol. Phys.*, 7 (1964) 211.
90 I. Tinoco, in B. Pullman and M. Weissbluth (Eds.), *Molecular Biophysics*, Academic Press, New York, 1965, p. 269.
91 D. F. Bradley, S. Lifson and B. Honig, in B. Pullman (Ed.), *Electronic Aspects of Biochemistry*, Academic Press, New York, 1964, p. 77.
92 L. Brillouin, in M. Kasha and B. Pullman (Eds.), *Horizons in Biochemistry*, Academic Press, New York, 1962, p. 295.
93 L. Salem, *Nature*, 193 (1962) 476.
94 L. Salem, *Can. J. Biochem. Physiol.*, 40 (1962) 1287.
95 R. Rein and F. E. Harris, *J. Chem. Phys.*, 41 (1964) 3393.
96 A. Cheicucci, J. Depireux and J. Duchesne, *Compt. Rend.*, 259 (1964) 1669.
97 J. Marmur and P. Doty, *Nature*, 183 (1959) 1427.
98 J. D. Watson and F. H. C. Crick, *Nature*, 171 (1953) 964.
99 E. Freeze, in J. H. Taylor (Ed.), *Molecular Genetics*, Academic Press, New York, 1963, p. 207.
100 C. Hélène, A. Haug, M. Delbruck and P. Douzou, *Compt. Rend.*, 259 (1964) 3385.
101 C. Hélène and P. Douzou, *Compt. Rend.*, 258 (1964) 196; 259 (1964) 4853.
102 K. Denbigh, *Trans. Faraday Soc.*, 36 (1940) 936.
103 H. DeVoe and I. Tinoco Jr., *J. Mol. Biol.*, 4 (1962) 500.
104 B. Pullman, P. Claverie and J. Caillet, *Proc. Natl. Acad. Sci. (U.S.)*, 55 (1966) 904.
105 P. D. Lawley and P. Brookes, *Biochem. J.*, 89 (1963) 127; *Exptl. Cell Res.*, 9 (1963) 512.
106 P. Brookes, in P. A. Plattner (Ed.), *Chemotherapy of Cancer*, Elsevier, Amsterdam, 1964, p. 32.
107 R. S. Mulliken, *J. Chim. Phys.*, 61 (1964) 20.
108 E. M. Kosower, in P. O. Boyer, H. Lardy and K. Myrbäck (Eds.), *The Enzymes*, Vol. 3, Academic Press, New York, 1960, p. 171.
109 G. Cilento and P. Giusti, *J. Am. Chem. Soc.*, 81 (1959) 3801.
110 S. G. A. Alivisatos, G. A. Mourkides and A. Sibril, *Nature*, 186 (1960) 718.
111 S. G. A. Alivisatos, F. Ungar, H. Sibril and G. A. Mourkides, *Biochim. Biophys. Acta*, 51 (1961) 361.

112 S. SHIFRIN, *Biochemistry*, 3 (1964) 829.
113 S. SHIFRIN, *Biochim. Biophys. Acta*, 81 (1964) 205.
114 G. CILENTO AND S. SCHREIER, *Arch. Biochem. Biophys.*, 107 (1964) 102.
115 D. A. KEARNS AND M. CALVIN, *J. Chem. Phys.*, 34 (1961) 2026.
116 L. E. LYONS AND J. C. MACKIE, *Nature*, 197 (1963) 589.
117 P. J. ELVING, W. A. STRUCK AND D. L. SMITH, *Chim. Anal. Org. Pharm. Chromatog.*, 14 (1965) 141.
118 D. S. SMITH AND P. J. ELVING, *J. Am. Chem. Soc.*, 84 (1962) 2741.
119 B. JANIK AND E. PALECEK, *Arch. Biochem. Biophys.*, 105 (1964) 225.
120 W. A. STRUCK AND P. J. ELVING, *J. Am. Chem. Soc.*, 86 (1964) 1229.
121 C. LAGERCRANTZ AND M. YHLAND, *Acta Chem. Scand.*, 17 (1963) 904, 1677.
122 Y. C. ORR, *Nature*, 201 (1964) 816.
123 S. MATSUHITA, F. IBUKI AND A. AOKI, *Arch. Biochem. Biophys.*, 102 (1963) 446.
124 T. GLAVIND AND E. SONDERGAARD, *Acta Chem. Scand.*, 18 (1964) 2173.
125 J. DUCHESNE, P. MACHMER AND M. READ, *Compt. Rend.*, 260 (1965) 2081.
126 P. MACHMER AND J. DUCHESNE, *Nature*, 206 (1965) 618.
127 B. PULLMAN, P. CLAVERIE AND J. CAILLET, *Science*, 147 (1965) 1305.
128 J. C. M. TSIBRIS, D. B. McCORMICK AND L. D. WRIGHT, *Biochemistry*, 4 (1965) 504.
129 A. VAN DER WORST AND A. PULLMAN, *Compt. Rend.*, 261 (1965) 827.
130 A. SZENT-GYORGYI, *Nature*, 148 (1941) 157.
131 D. D. ELEY, in M. KASHA AND B. PULLMAN (Eds.), *Horizons in Biochemistry*, Academic Press, New York, 1962, p. 341.
132 D. D. ELEY AND R. B. LESLIE, in B. PULLMAN (Ed.), *Electronic Aspects of Biochemistry*, Academic Press, New York, 1964, p. 105.
133 B. ROSENBERG, *Biopolymers Symp.*, 1 (1964) 453.
134 B. ROSENBERG, in L. AUGENSTEIN, R. MASON AND B. ROSENBERG (Eds.), *Physical Processes in Radiation Biology*, Academic Press, New York, 1964, p. 111.
135 J. DUCHESNE, in M. KASHA AND B. PULLMAN (Eds.), *Horizons in Biochemistry*, Academic Press, New York, 1962, p. 335.
136 C. T. O'KONSKI, P. MOSER AND M. SHIRAI, *Biopolymers Symp.*, 1 (1964) 479.
137 D. R. KEARNS, in J. DUCHESNE (Ed.), *The Structure and Properties of Biomolecules and Biological Systems*, Interscience, New York, 1964, p. 282.
138 E. E. SNELL, in *Ciba Foundation Symposium on the Mechanism of Action of Water-Soluble Vitamins*, Churchill, London, 1961, p. 20.
139 A. PULLMAN AND B. PULLMAN, *Advan. Cancer Res.*, 3 (1955) 117.
140 A. PULLMAN AND B. PULLMAN, *Cancerisation par les Substances Chimiques et Structure Moleculaire*, Masson, Paris, 1955.
141 E. C. MILLER AND J. A. MILLER *Cancer Res.* 11 (1951) 100; 12 (1952) 547.
142 J. A. MILLER AND E. C. MILLER, *Can. Cancer Conf.*, 4 (1961) 57.
143 C. HEIDELBERGER, *Acta Unio Intern. Contre Cancrum*, 15 (1959) 107.
144 P. M. BHARGAVA, H. I. HADLER AND C. HEIDELBERGER, *J. Am. Chem. Soc.*, 77 (1955) 2877.
145 P. M. BHARGAVA AND C. HEIDELBERGER, *J. Am. Chem. Soc.*, 78 (1956) 3671.
146 C. HEIDELBERGER AND M. G. MOLDENHAUER, *Cancer Res.*, 16 (1956) 442.
147 V. T. OLIVERIO AND C. HEIDELBERGER, *Cancer Res.*, 18 (1958) 1094.
148 C. HEIDELBERGER, *J. Cellular Comp. Physiol.*, 64, Suppl. 1 (1964) 129.
149 P. BROOKES AND P. D. LAWLEY, *Nature*, 207 (1964) 781.
150 R. REIN AND F. E. HARRIS, *J. Chem. Phys.*, 41 (1964) 3393; 42 (1965) 2177; 43 (1965) 4415.
151 P. O. LÖWDIN, *Advan. Quantum Chem.*, 2 (1965) 213.

152 A. Pullman and B. Pullman, in P. O. Löwdin (Ed.), *The Quantum Theory of Atoms, Molecules and Solid State*, Academic Press, New York, 1966.
153 B. Pullman and M. J. Mantione, *Compt. Rend.*, 260 (1965) 5643; 261 (1965) 5679.
154 J. P. Green and J. P. Malrieu, *Proc. Natl. Acad. Sci. (U.S.)*, 54 (1965) 659.
155 W. B. Neely, *Mol. Pharmacol.*, 1 (1965) 137.
156 D. Agin, L. Hersch and D. Holtzman, *Proc. Natl. Acad. Sci. (U.S.)*, 53 (1965) 952.
157 M. Cocordano and J. Ricard, *Physiol. Vég.*, 1 (1963) 129.
158 T. Ban, *Japan. J. Pharmacol.*, 12 (1962) 72.
159 H. Berthod, C. Giessner-Prettre and A. Pullman, *Theoret. Chim. Acta*, 5 (1966) 53.

Chapter II

Mechanisms of Energy Transfer

TH. FÖRSTER

Laboratory for Physical Chemistry, Technical University, Stuttgart (Germany)

1. Introduction

There is evidence from different sources that many biological systems are able to transfer energy, or, more specifically: free energy, between different components, even if these are not in spatial contact with each other[1]. Most of this evidence has been derived from experiments in photo- and radio-biology. The ability to transfer energy is even widely believed to be a general property of biological material and to be intimately related to biological functions.

During the last few decades, different mechanisms have been proposed for this transfer. A simple mechanical transport by the flow or the diffusion of radicals or of energy-rich molecules may be responsible in some cases but can be excluded in others. Energy transport by mobile electrons[2,3] has frequently been discussed, but the evidence is rather against conductive or semiconductive properties of the material. Proton transfer[4] by some kind of Grotthus mechanism might be possible over some distance but should not be able to transfer much energy. Therefore, there is still much interest in other mechanisms, by which free energy, possibly without any material carrier, might be transferred between separate entities within a macromolecular structure.

We have specifically mentioned free energy, because under the prevailing isothermal conditions only this is useful energy which is not dissipated. In isothermal chemical systems, free energy is exchanged by the transformation of electronic energy such as present in chemical bonds. In photo- and radio-

References p. 78

biology, electronically excited states, either singlet- or triplet states, participate in the transformation processes. Charge-transfer complexes are common
constituents of biological systems[5]. They have excited electronic states of
comparatively low energy, in which free energy might be stored and submitted
in transfer processes.

It has been demonstrated by experiment and explained by theory that
electronic excitation energy can be transferred, without essential degradation,
by a non-radiative process over considerable distances[6]. Under suitable
conditions this transfer occurs over distances up to 50 Å or more in one single
step. By successive steps that transfer can pass a considerable number of
molecules and proceed over even larger distances. In biological materials the
conditions for this are quite favourable because of the fairly high local
concentrations of components with low excitation energy, such as aromatic
amino acids and pigments. It is a general experience that electronic excitation
of these constituents, if it is produced by the absorption of radiation, is
transferred from one to another. Such transfer processes are now regarded
as essential steps of photo- and radiobiological reactions[7].

Electronic excitation and subsequent transfer processes should be expected
even in biological dark processes. One might argue that the energy quanta
available in, for instance, enzymatic processes are too low for the electronic
excitation of aromatic amino acids, and that pigments with sufficiently low-
lying excited states are not generally present. Nevertheless, as it has already
been mentioned, electronic excitations may occur in charge-transfer complexes. Such complexes consist of two components, one (A) with low ionization energy and the second (B) with high electron affinity. From the normal
configuration AB of the complex the configuration A^+B^- results by the shift
of one electron. This configuration is an excited electronic state of the complex in the same sense as are other excited states in single molecules. It might
be of low enough energy to become excited by moderately exothermic
reactions. Although this has not yet been demonstrated, energy transfer
should be expected between different complexes of that kind, even if they are
so far apart that electrons cannot be transferred from one to another, but
within the separate complexes only. In chemical language, this might be
considered as an energetic coupling between separated redox systems.

Whether or not the speculations of the foregoing paragraph are permissible,
the rôle of excitation transfer in photo- and radiobiology justifies the consideration of intermolecular excitation transfer as a basic mechanism. With
these applications in mind, the term molecule shall be used here and later on

in a more general sense to include molecular complexes and separate components within larger macromolecular structures.

2. Theory of excitation transfer

(a) Classifications

Excitation transfer may involve different kinds of electronic states and may be accomplished by different interaction mechanisms. Besides this, it proceeds quite differently depending on the magnitude of the interaction, but regardless of its nature. These differences may be a basis for a classification of excitation-transfer processes[8-10].

The important excited states of stable molecules are either short-lived *singlet*-(S-)states or longer living *triplet*-(T-)states, both of which can be involved in the transfer. We disregard charge-transfer processes, because these will be considered in another contribution to this volume[5]. An elementary transfer process between two molecules consists, then, in the deactivation of the first, the donor molecule, coupled to an activation in the other, the acceptor molecule. The following processes may be considered as examples:

$$1 \quad S_1 + S_0' \rightarrow S_0 + S_1' \qquad \text{(singlet} \rightarrow \text{triplet)}$$

$$2 \quad T_1 + S_0' \rightarrow S_0 + S_1' \qquad \text{(triplet} \rightarrow \text{singlet)}$$

$$3 \quad S_1 + T_1' \rightarrow S_0 + T_2' \qquad \text{(singlet} \rightarrow \text{higher triplet)}$$

$$4 \quad T_1 + S_0' \rightarrow S_0 + T_1' \qquad \text{(triplet} \rightarrow \text{triplet)}$$

Here, the "prime" designates the acceptor and the subscript the electronic state, with zero for ground state. All these processes occur, of course, as *resonance* processes, that is, under conservation of total energy. This implies that the electronic energy of the acceptor state has to be either the same or less than that of the donor state. A surplus of electronic energy may be transformed into vibrational energy or, *vice versa*, a small deficiency may be covered by available thermal energy. In systems containing many molecules with suitable excitation energies multistep processes may occur by such elementary processes in sequence.

Under otherwise comparable conditions, *spin conservation*, or nonconservation, is the most essential rate-determining factor. In the processes 1, 3 and 4, the total spin of the system is unchanged but in 2 it is changed. Nevertheless the latter process is possible, because all electronic states are impure spin states for which spin conservation is not strictly obeyed. In processes 1 and 3, even the individual spins of donor and of acceptor remain unchanged. The *efficiency* of a transfer process depends on the ratio of the transfer rate to that of the intramolecular deactivation in the donor. This process, too, is essentially governed by spin rules. Therefore, spin conservation in the donor is less important, but spin conservation in the acceptor is the decisive factor which, under otherwise comparable conditions, determines the transfer efficiency[8]. This makes the processes 1, 2 and 3 occur under less stringent conditions than process 4.

As stated before, excitation transfer can result from different interaction mechanisms. The interaction may be (either) Coulomb or exchange interaction. The first one can be regarded as the quasi-electrostatic interaction between the appropriate transition-charge densities of the intramolecular transitions involved in the transfer (*e.g.* $S_1 \leftrightarrow S_0$ and $S'_1 \leftrightarrow S'_0$ in process 1). This interaction predominates, if the transition is spin- and symmetry-allowed in the acceptor. Its expression for large distances has then a leading dipole–dipole term with R^{-3} distance dependence which may give rise to long-range transfer. For symmetry forbidden transitions higher multipole terms or vibrationally induced dipole terms may be important. For spin-forbidden transitions in the acceptor exchange interaction predominates[13]. As this requires some overlap between the electronic clouds of both molecules, it decreases very steeply with the distance and allows, therefore, only short-range transfer in a single step. It is the main source of interaction in process 4, whereas, in the others, Coulomb interaction usually predominates. Occasionally, charge-transfer interaction, which is not considered here, may contribute to the other sources.

(b) Strong coupling

The magnitude of the interaction is even more important than its nature. It leads to a fairly clean distinction between three main *coupling cases*[10] which now are commonly called those of strong, weak and very weak coupling[11,12]. The coupling is *strong* if the intermolecular interaction exceeds that between the electronic and the nuclear motions within the individual molecules. Under such strong coupling all the vibronic subtransitions in both molecules are virtually at resonance with each other. The transfer of excitation is then faster than nuclear vibrations and occurs without any essential readjustment of the nuclear equilibrium positions. It is adequate here to consider the electronic excitation not as temporarily localized on one or the other molecule, but as distributed over all of them. This corresponds to a description of the system by its stationary excited electronic states, which have been defined and named *exciton* states by Frenkel[14]. Transfer processes are, then, described by non-stationary, temporarily localized states which can be composed from these *exciton* states. For two molecules in resonance with each other, that is, with their excitation energies differing by less than the interaction energy U, the transfer rate is approximately[10]

$$n^T \sim \frac{4|U|}{h} \tag{1}$$

in which h = Planck's constant, n^T = rate of transfer.

In multicomponent systems, such as one- or more-dimensional lattices, successive transfer processes are coherent. The exciton migrates straight forward with constant group velocity, that is, with the r.m.s. (root-mean-square)-average distance from the original site of the excitation in proportion to time.

Between nearest neighbours the transfer rate is then[15,16] essentially that in Eqn. 1 with a numerical factor depending on the lattice type. The migration is not much disturbed by smaller irregularities in the lattice structure. Stronger irregularities lead to a decrease in coherence so that the propagation finally becomes more diffusive, with the r.m.s.-average distance proportional to the square root of time[17]. In this context, medium-transmitted interaction should be mentioned[18]. This may occur by some kind of a virtual exciton interaction between a pair of molecules which are indirectly coupled to each other by a chain of molecules with their excitation energies far out of resonance with the first ones.

Phenomenologically, strongly coupled systems are characterized by large differences between their absorption spectra and those of their components. These differences appear even in the vibronic envelopes and can thus be recognized even from unstructured spectra.

(c) Weak coupling

If the intermolecular interaction is less than the intramolecular one between electronic and nuclear motion, the coupling is called *weak*. The distinction between this and the foregoing coupling case was recognized quite early[19] but has especially been emphasized later by Simpson and Peterson[20]. Both cases and their intermediates have been extensively discussed[20-25]. Under weak coupling the condition for resonance is more stringent than for strong coupling, so that even in molecules of the same kind, only *pairs* of vibronic transitions are at resonance with each other[8]. Excitation transfer then needs several vibrational periods, and it is connected with the readjustment of the nuclear equilibrium positions as well as with a transfer of vibrational excitation. It is evident that the electronic excitation is here more localized than under strong coupling. Nevertheless it is now adequate to consider the vibronic excitation still as delocalized and to describe the system by its stationary *vibronic* exciton states.

In a weakly coupled dimer, with the vibrational quantum numbers v and w of the excited and of the unexcited component, respectively, the transfer rate is now in the simplest case[10,12]

$$n_{v,w}^T \sim \frac{4|U|S_{vw}^2}{h} \tag{2}$$

Here, $S_{vw} < 1$ is the Franck–Condon integral of the intramolecular transition $v \leftrightarrow w$. Therefore, the transfer rate is less than for strong coupling, even with the same value of U. The product US_{vw}^2 in Eqn. 2 may be regarded as the interaction energy between the vibronic transitions involved in the process.

In an ideal lattice the vibronic excitation migrates slower than an electronic exciton under strong coupling, but still with constant group velocity. Because of the sharper resonance condition, however, the coherent propagation is more sensitive to lattice irregularities and even lattice vibrations. This leads to scattering processes which restrict the free path length of the exciton to a few or more lattice distances, the number of which might be estimated from the ratio of the vibronic interaction energy to the vibronic band width. Over more than the free-path-length distances, the propagation becomes incoherent or diffusive[26].

As one might expect, weak coupling leads to minor alterations in the absorption spectrum. The vibrational envelope of the single molecule is essentially retained and differences exist in the individual vibronic bands only. In typical examples these show characteristic splitting patterns with intervals of the magnitude 2 US_{vw}^2. This is the so-called (vibronic) *Davidov* splitting, which is commonly observed in the spectra of crystals with more than one molecule in the unit cell[27]. Obviously, this phenomenological characteristic of weak coupling is restricted to systems with distinct vibrational structure. A detailed consideration leads to the important result that weak coupling exists only in these systems[12].

(d) Very weak coupling

With still less coupling energy the condition for resonance becomes more and more stringent. Finally, the finite width $\Delta\varepsilon$ of the vibronic bands becomes essential. If the vibronic interaction energy US_{vw}^2 is less than $\Delta\varepsilon$, the properties of the system differ essentially from those of a weakly coupled system. We now have the third of our coupling cases, that of *very weak* coupling. Roughly speaking, this can be regarded as a case in which even from two coinciding vibronic bands certain regions only are at resonance with each other[10,12]. This introduces into the expression for the transfer rate an additional factor of the order $S_{vw}^2/\Delta\varepsilon$. According to more accurate calculations the very weak coupling transfer rate is

$$n_{v,w}^T \sim \frac{4\pi^2 \, U^2 \, S_{vw}^4}{h \, \Delta\varepsilon} \tag{3}$$

This refers to an initial condition in which both molecules are in well-defined vibrational levels v and w. For an original distribution over such levels the total transfer rate n_{tot}^T becomes a sum over the individual transfer rates n_{vw}^T, each of them multiplied by the appropriate distribution functions.

According to Eqn. 3, the very weak coupling transfer rate is proportional to the square of the vibronic interaction energy. For dipole–dipole interaction this leads to an inverse 6th power distance dependence of the transfer rate[28,29]. The transfer is then often said to occur by "inductive resonance".

The finite width of an individual vibronic transition may be interpreted as resulting from life-time broadening. This is due to the redistribution of vibrational energies within the separate molecules, under assistance of external degrees of freedom of the molecules involved in transfer and of others. It is nothing else than thermal relaxation which in condensed systems restricts the lifetimes of individual vibronic states to 10^{-12} sec or less at room temperature. Under very weak coupling conditions an excited molecule not only performs several vibrations but also occupies different vibronic levels before any transfer occurs. Here we prefer to regard the excitation as temporarily localized on one single molecule*. It is a matter of taste, whether to call this a localized exciton or simply an excitation.

Very weak coupling does not lead to any recognizable difference between the absorption spectrum of the system and that of its components. In spectra with vibronic structure all imaginable band splittings are so small that they are obscured by the diffuseness of the bands. The same holds even more for spectra without essential vibronic structure such as are quite common for more complicated molecules in solution. For these, either alterations of the vibronic envelope or no alterations at all can be recognized. This corresponds to either strong or very weak coupling and, as mentioned before, weak coupling does not exist for such molecules[12].

Under very weak coupling conditions, the excitation propagates in a multimolecular system simply by a sequence of the uncorrelated individual transfer processes. The propagation is now completely diffusive, with a free-path length equal to the distance between nearest neighbours. In a lattice the diffusion constant is obtained as the square of this distance divided by the corresponding transfer time and multiplied by a numerical factor depending on the lattice type[30]. Diffusive propagation is the characteristic feature of the so-called "hopping model"[32-34] which would seem to be justified only in the very weak coupling case.

(e) Long range dipole–dipole transfer

With some additional assumptions the transfer rate for this can be expressed more explicitly. If the transfer is so slow that it occurs between thermally equilibrated molecules, the total transfer rate n_{tot}^T contains certain combinations of vibrational distribution functions and Franck–Condon factors. These are exactly the same combinations which determine the intensities of the vibronic transitions in the emission spectrum of the donor and in the absorption spectrum of the acceptor. Under the further assumption of pure dipole–dipole interaction this leads to the following expression, which can be obtained either by quantum-mechanical theory[12,31] or even from classical arguments[6,35].

---

* In the author's opinion, the conceptions of delocalized excitation and of excitation transfer are complementary but the second one is more appropriate to very weak coupling conditions. A more penetrating investigation has led Bay and Pearlstein[30] to the result that even in very weakly coupled systems the excitation should better be described as delocalized. This does not, however, lead to different predictions for experimental results.

$$n^T = \frac{9\kappa^2 (\ln 10)}{128 \pi^5 n^4 N' \tau_d^i} \frac{1}{R^6} \int f_d(\tilde{v}) \varepsilon_a(\tilde{v}) \frac{d\tilde{v}}{\tilde{v}^4} \tag{4}$$

Here $\tilde{v}$ is the wave number, $\varepsilon_a(\tilde{v})$ the molar decadic extinction coefficient of the acceptor and $f_d(\tilde{v})$ the spectrum of the donor emission (fluorescence or phosphorescence) measured in quanta per wave-number interval and normalized to unity on the same scale. N' is the number of molecules per millimole, n the refractive index of the surrounding medium and τ_d^i the intrinsic life time of the excited donor state. κ is a numerical factor representing the orientation dependence of dipole–dipole interaction. Its average is $\sqrt{2/3} = 0.816$ for fast Brownian rotation of both molecules[31] and 0.690 for random but rigid orientations[36].

Eqn. 4 gives the very weak coupling transfer rate for exclusive dipole–dipole interaction under thermal equilibrium. It is valid regardless of the nature of states involved, whether the molecules are of the same or of different kind, and whether their spectra have vibrational structure or not. Apart from the factor $\tilde{v}^{-4}$ the integral represents the overlap of the emission spectrum of the donor with the absorption spectrum of the acceptor. Obviously, an acceptor with a little less electronic energy than that of the donor is especially favourable here. Nevertheless, transfer between equivalent states of alike molecules is not so much slower at room temperature where emission and absorption spectra of the same molecule overlap considerably. Excitation transfer under the conditions of Eqn. 4 is often called long-range dipole–dipole transfer, because it is the only non-radiative type of transfer which, in a single step, occurs over much more than molecular dimensions.

For practical purposes Eqn. 4 may be written

$$n^T = \frac{1}{\tau_d} \left(\frac{R_0}{R} \right)^6 \tag{5}$$

where τ_d is the actual donor state lifetime and R_0 a critical transfer distance, at which the rates of transfer and of spontaneous deactivation of the donor are equal. One obtains then

$$R_0^6 = \frac{9\kappa^2 (\ln 10) \eta_d}{128\pi^5 n^4 N'} \int f_d(\tilde{v}) \varepsilon_a(\tilde{v}) \frac{d\tilde{v}}{\tilde{v}^4} \tag{6}$$

where $\eta_d = \tau_d / \tau_d^i$ is the quantum efficiency of the donor emission (for dipole radiation) in absence of the acceptor. For $\eta_d \sim 1$, with a strongly absorbing acceptor and optimal spectral overlap, R_0 can attain values between 50 and 100 Å. It should be noted that R_0 does not depend on the lifetime of the donor state, but on the quantum efficiency of its emission. Therefore, long-range dipole–dipole transfer does not necessarily require a donor singlet state, but may occur as well from donor triplet states[8] such as in process d (p. 74). The lower rate of the transfer process is, then, compensated by the longer lifetime of the transferring state.

The very weak coupling formulation has been generalized by Dexter[13] for higher multipole and exchange interactions. It is obvious that, in these cases, the transfer rate depends more critically on the intermolecular distance and becomes significant only for short intervals.

Experimental information on transfer under very weak coupling is usually gained from solutions where not only the orientations but also the distances between donor and acceptor molecules are statistical. If transfer between different donor molecules is negligible, the transfer efficiency η^T is then a function of the acceptor concentration alone. For dipole–dipole interaction between rigidly located molecules this is[37,38]:

$$\eta^T = 2x e^{x^2} \int_0^\infty e^{-t^2} \, dt = \sqrt{\pi} x e^{x^2} (1 - \Phi(x)) \tag{7}{}^*$$

where $\Phi(x)$ is the error function, $x = c/c_0$ and

$$c_0 = \frac{3}{2\sqrt{\pi^3}} \frac{1}{N' R_0^3} = \left(\frac{7.66 \cdot 10^{-6}}{R_0}\right)^3 \tag{8}$$

c_0 may be regarded as a critical transfer concentration**, corresponding to a transfer efficiency of 76%. In the presence of the acceptor, not only the quantum efficiency but also the duration of the donor emission is reduced, but less than the former. Furthermore, the decay is non-exponential and depends on the duration of the excitation, because transfer occurs preferentially between close lying donor–acceptor pairs. The resulting decay laws have been formulated for the two extreme conditions of long- and of short-exciting pulses[43].

Recently, the theory of long-range dipole–dipole transfer has been extended

---

* The derivation of this expression has given rise to some controversies[39,40] which, how-ever, do not invalidate the result[41,42].

** It should be noted that different definitions of c_0 are used in the literature[31]. The present one is preferable for statistically distributed molecules.

further to include Brownian translation in nonrigid solvents[44-46]. Similar formulas have also been given for exchange transfer in solutions with statistically distributed molecules[47].

3. Experimental investigations

(a) Methods

It is not intended here to consider excitation transfer in biological systems. Instead of this, a few examples of delocalized excitation and of excitation transfer in solutions and crystals will be given with the intention of illustrating the theoretical developments of the foregoing sections and to demonstrate the reality of the effects treated there.

Experimental evidence on excitation transfer derives from different sources. In strongly and weakly coupled systems most of the information has been obtained from absorption spectra and is, thus, rather on the delocalization of excitation. In weakly coupled systems the direct observation of actual excitation transfer is the only source. Depending on the nature of excited states involved, different effects are to be used here, such as fluorescence, phosphorescence, electron-spin resonance and even chemical reactivity.

(b) Delocalization

Typical examples for strong coupling are the dimeric aggregates which exist in more concentrated aqueous solutions of many dyes such as thionine or the rhodamines[48-50]. Their absorption spectra are much broader than those of the monomers and, in several cases, consist of two distinct components with peak-to-peak separations of 1000 cm^{-1} or more, and comparable to the monomer band width. Similar but somewhat smaller splittings occur in the spectra of some double molecules, such as the p-cyclophanes[51] and hydrogen-bonded dimers[51a].

The most illuminating examples are the polymers present in concentrated aqueous solutions of pseudo-isocyanine[52,53] and similar dyes. Their absorption spectra culminate in one sharp band which is interpreted as resulting from the transition to the edge of the exciton band of the polymer. The unusual sharpness of that band indicates the essential reduction of vibrational excitation, which occurs under electronic delocalization in a multicomponent system. In the more intense absorption regions of some molecular crystals, strong coupling behaviour has been observed too. The first electronic transition of perylene[54] and the second one of anthracene[55] may be mentioned as examples.

In most organic crystals, however, the first absorption regions are less intense and show weak coupling characteristics. In the anthracene crystal, for instance, the vibronic envelope of that transition is nearly the same as in solution. However, the vibronic bands are individually split, each into two components, with separations[56,57] from 100–200 cm^{-1}. With some reservations[58,59] we may say that the vibronic interaction energies between nearest neighbours in the lattice are of the same order of magnitude.

(c) Single-step transfer

Direct information on excitation transfer is obtained from sensitization experiments in which the transfer results in the replacement of donor-emission by acceptor-emission. The first demonstration of fluorescence sensitization was achieved with atomic vapours by Cario and Franck[60] in 1922. Some years later, Perrin and Choucroun (1927)[61] reported similar experiments on singlet–singlet transfer in solution by the process

$$S_1 + S_0' \rightarrow S_0 + S_1'$$

Quantitative investigations on fluorescence sensitization in solution and their comparison with theory began in 1949. One of the most studied systems is that with trypaflavin as the donor and rhodamine B as the acceptor[37,62-68]. For this system the validity of Eqn. 7 was established and the critical transfer distances R_0 (between 54 and 58 Å) were found to agree quite well with its theoretical value derived from Eqn. 6. With this and many other systems further characteristics of long range dipole–dipole transfer were established, such as the concentration dependence of the decay of donor- and acceptor-emission[67,68], the increase in the total fluorescence efficiency by transfer from a donor with low quantum efficiency to an acceptor with a higher one[69,70], the correlation[71] with the spectral overlap integral of Eqn. 6, and the validity of an expression analogous to that in Eqn. 7 for a two-dimensional solution[72]. Recently, a careful study of the non-exponential decay of the donor fluorescence has verified exactly the R^{-6}-distance dependence of long range dipole–dipole transfer[73].

It may be regarded as a handicap of such experiments with random distributions of donor and acceptor molecules that they refer to an average of transfer processes over different distances. The demonstration of transfer over a well-defined donor–acceptor distance would seem more conclusive. For this, something like a "molecular stick" is required in order to hold the components apart at a distance of the order R_0. After earlier attempts[74] this has now been realized by a rigid bisteroid framework to which donor and acceptor are condensed and held[75] at a distance of about 20 Å. The donor and acceptor groups used here were p-methoxyphenyl/1-naphthoyl and 1-naphthoyl-9-anthracenecarboxyl. With the first pair exact, with the second a fair agreement was found between the experimental R_0 values and those calculated from spectral data.

Some other experiments by the Marburg group should be mentioned here[76,77]. They might be named "molecular sheet" experiments. Here, mono-molecular layers of two different dyestuffs, one the donor and the other the acceptor, were separated by multiple layers of an inert surface-active material, 50 to several 100 Å thick. The transfer efficiency from one side of that sandwich to the other was found to decrease with the inverse 4th power of the distance. For two-dimensional layers this corresponds to a R^{-6}-distance dependence of the elementary transfer process. It must be mentioned, however, that the experimental R_0 values are essentially higher than calculat-ed. This might result from the mutual interaction between the close-lying dyestuff molecules in their individual layers.

In the singlet–singlet transfer processes discussed here, the intramolecular transitions are allowed in the donor and in the acceptor as well. The same situation prevails in the singlet–higher-triplet transfer process

$$S_1 + T_1' \rightarrow S_0 + T_2'$$

This process requires a somewhat unusual experimental condition with the acceptor already present in its triplet state. Nevertheless, with perylene as the donor and phenanthrene-d_{10} the acceptor in rigid solution, this process has been established to occur as long-range transfer[78].

As stated earlier, this kind of transfer may even occur if the transition in the donor is spin-forbidden, because the slower transfer rate can be compensated by the longer lifetime of the donor state. Accordingly, the triplet–singlet transfer process

$$T_1 + S_0' \rightarrow S_0 + S_1'$$

should occur[8] with an acceptor of a sufficiently low-lying excited singlet and give rise to delayed fluorescence of the acceptor under quenching of the donor phosphorescence. This has been observed by Ermolaev and Sveshni-kova[79,80], and later experiments established the R^{-6}-distance dependence of the process[81].

The two foregoing processes are related to another process

$$T_1 + T_1 \rightarrow S_0 + T_2$$

which should be of importance for long-range triplet–triplet annihilation in rigid inert solutions[81].

The common feature of all the aforementioned processes is the spin-allowedness in the acceptor, which permits them to occur as long-range processes in one single step by dipole–dipole interaction under very weak coupling conditions.

For a spin-forbidden transition in the acceptor, the Coulomb interaction energy is not sufficient for transfer during the lifetime of the excited donor state. Exchange interaction prevails here. The most common example is triplet–triplet transfer

$$T_1 + S_0' \rightarrow S_0 + T_1'$$

This process has been discovered by Terenin and Ermolaev[82,83] and has been extensively studied with suitable systems in rigid glasses. The critical transfer distances are comparatively short, between 10 and 15 Å, and the dependences of transfer efficiencies and decay rates on the acceptor concentration indicate a very steep distance dependence. This process has also been studied in less rigid and fluid solutions, where it is diffusion-controlled[84–86].

Excitation-transfer processes involving triplet states have been studied by various other methods, such as flash spectroscopy[87,88] and electron-spin resonance[89–93]. With the latter method the important result was obtained[94] that the glassy solvent does not participate in triplet–triplet transfer, as long as the triplet states of the solvent molecules are not accessible from the donor triplet by thermal activation.

(d) Multi-step transfer

Multi-step transfer occurs by successive elementary single-step processes with or without correlation between them. It requires a high density of molecules of the same kind, such as present in more concentrated solutions or in crystals. The so-called concentration depolarisation of fluorescence[95] in isotropic solutions has been explained by such repeated step singlet–singlet transfer between differently oriented molecules. In typical transfer experiments, another molecular species is present at lower concentration, in order to act as the final acceptor. A peculiar type of concentration quenching solution has been interpreted by such a sequence of transfer processes between monomer molecules, with termination at one of a few quenching dimers[96,97]. In both examples the coupling is very weak, so that the different steps are uncorrelated with each other.

Fluorescence-sensitization experiments have been made with molecular crystals containing a small amount of an impurity with lower excitation energy. The best known example is that of anthracene crystals with traces of naphthacene[98,99]. These crystals emit the naphthacene fluorescence even at an impurity mole ratio of 10^{-5} with exclusive excitation of the host molecules. Here the multi-step nature of the transfer process is evident because of the weak coupling properties of the host material. It is further substantiated by the low concentration of the acceptor required and by the decrease in transfer efficiency at low temperature[100,101]. The transfer is expected here to occur by the migration of vibronic excitons, which, after more or less efficient scattering, are finally trapped by an impurity molecule with its lower energy state.

This and many similar crystalline systems have been studied quantitatively[102]. Unfortunately, the participation at the overall process of two different elementary processes (host-to-host and host-to-guest) complicates the evaluation of these experiments. Most authors satisfy themselves with a simplified, so called "hopping model"[34], by which the excitation is supposed to jump between neighbouring host molecules, until it arrives at a guest molecule. This model neglects the coherence between successive elementary steps and is therefore, not exactly in accord with the weak-coupling properties of the host lattice.

Recently, multi-step triplet–triplet transfer processes have been found to play an essential rôle in molecular crystals. They occur mainly by exchange interaction which is sufficiently strong with the close distances between adjacent molecules in the lattice. Charge-transfer interaction may be a supplementary mechanism[103]. Experimental and theoretical estimates lead to a magnitude of vibronic interaction of about 10 cm^{-1}. This makes the rate of the individual transfer process less than that between singlet states under otherwise comparable conditions. It permits, however, even more extensive migration of triplet excitation during the much longer triplet state lifetime. Phosphorescence emission from impurities, sensitized by host excitation, is thus a common property of molecular crystals.

The mobility of triplet excitation, together with its long intrinsic lifetime, makes this excitation extremely sensitive against spurious quenching impurities. Therefore, most supposedly "pure" organic molecular crystals are only weakly phosphorescent. In addition to impurity quenching, biphotonic triplet–triplet annihilation occurs even at moderate intensities of excitation[104]. This annihilation leads partly to the formation of an excited singlet again ($T_1 + T_1 \rightarrow S_1 + S_0$) which returns to the ground state under emission of delayed fluorescence.

Because of the smaller interaction energy the propagation of triplet excitation is less coherent than that of singlet excitons. With the exception of very low temperatures, the possible exciton splitting is certainly less than the bandwidth, so that the very weak coupling theory and the hopping model should apply. From delayed fluorescence experiments a diffusion coefficient of $5 \cdot 10^{-4}$ cm$^2 \cdot$ sec^{-1} and a diffusion length of about 30 μ have been estimated for anthracene crystals, but a more direct experiment[106,107] leads to a value of about 10 μ. Even the smaller of these figures demonstrates the extensive mobility of triplet excitation in molecular crystals.

4. Applications to biological systems

The density of components with low excitation energies, such as aromatic amino acids, pigments and possibly charge-transfer complexes, favours excitation transfer in those systems. The structural regularities of polypeptides and polynucleotides, for instance, might even lead to the expectation of properties similar to those present in molecular crystals. It is the author's opinion, however, that this analogy should not be stressed too much. The regularities of biological structures are certainly much less than those of the crystals of typical aromatic compounds and, if strict translational symmetry is present at all, the identity period is much larger.

Furthermore, the individual aromatic amino acids and other components of biological systems are of a much more complicated structure than simple aromatic hydrocarbons in typical molecular crystals. They possess either more vibrational degrees of freedom or are, at least, less symmetric. As a consequence of this, electronic excitation is accompanied by the excitation of a greater number of fundamental frequencies. The numerous individual vibronic bands overlap each other extensively, so that the spectra are either continuous or possess only crude vibrational structure*. It has been stated in

---

* Such crude structure is present in the spectra of plant pigments, but the spectra of chloroplasts do not show the features of coupling, either strong or weak.

section 2d (p. 68), that under these conditions no weak coupling case exists, so that the coupling has to be either strong or very weak.

Neither the spectra of polypeptides nor those of polynucleotides show the essential deviations from additivity in the vibrational envelopes of the component spectra which would be expected for strong coupling. This is not surprising, because even in molecular crystals with their more dense structures strong coupling is rarely found. In the present case, where weak coupling is excluded, excitation transfer should be governed by very-weak coupling theory, with the single-step transfer rate proportional to the square of the interaction energy and without coherence between successive steps. One should remember, however, that this does not necessarily imply the R^{-6}-distance dependence of dipole–dipole interaction but that other mechanisms (*e.g.* exchange interaction) are also compatible with very weak coupling.

The situation might be different at very low temperatures, where more complicated molecules such as the components of biological systems could possess spectra with definite vibronic structure. In this case, a transfer rate proportional to the first power of the interaction energy, as has been suggested recently[108], would be possible. This, however, seems unimportant, as long as we are not concerned with biological conditions in outer space!

REFERENCES

1 A. SZENT-GYÖRGYI, *Bioenergetics*, Academic Press, New York, 1957.
2 D. D. ELEY, in M. KASHA AND B. PULLMAN (Eds.), *Horizons in Biochemistry*, Academic Press, New York, 1962, p. 341.
3 S. I. LEACH, *Advan. Enzymol.*, 15 (1954) 1.
4 K. WIRTZ, Z. *Naturforsch.*, 3b (1947) 132.
5 F. J. BULLOCK, in M. FLORKIN AND E. H. STOTZ (Eds.), *Comprehensive Biochemistry*, Vol. 22, Elsevier, Amsterdam, 1967, p. 81.
6 TH. FÖRSTER, *Naturwiss.*, 33 (1946) 166.
7 L. M. N. DUYSENS, *Progr. Biophys. Biophys. Chem.*, 14 (1964) 40.
8 TH. FÖRSTER, *Discussions Faraday Soc.*, 27 (1959) 7.
9 TH. FÖRSTER, *Radiation Res.*, *Suppl.* 2 (1960) 326.
10 TH. FÖRSTER, in M. BURTON, I. S. KIRBY-SMITH AND I. L. MAGEE (Eds.), *Comparative Effects of Radiation*, Wiley, New York, 1960, p. 300.
11 M. KASHA, in L. G. AUGENSTEIN, R. MASON AND B. ROSENBERG (Eds.), *Physical Processes in Radiation Biology*, Academic Press, New York, 1964, p. 17.
12 TH. FÖRSTER, in O. SINANOĞLU (Ed.), *Modern Quantum Chemistry, Istanbul Lectures*, Vol. 3, Academic Press, New York, 1966, p. 93.
13 D. L. DEXTER, *J. Chem. Phys.*, 21 (1953) 836.
14 I. FRENKEL, *Phys. Rev.*, 37 (1931) 17, 1276.
15 R. E. MERRIFIELD, *J. Chem. Phys.*, 37 (1958) 647.
16 I. L. MAGEE AND F. FUNABASHI, *J. Chem. Phys.*, 34 (1961) 1715.
17 K. KATSUURA, *J. Chem. Phys.*, 40 (1964) 3527.
18 V. M. AGRANOVICH, *Opt. i Spektroskopiya*, 9 (1960) 113; *Opt. Spectry. (USSR)*, 9 (1960) 59.
19 J. FRANCK AND E. TELLER, *J. Chem. Phys.*, 6 (1938) 861.
20 W. T. SIMPSON AND D. L. PETERSON, *J. Chem. Phys.*, 26 (1957) 588.
20a D. S. McCLURE, *Can. J. Chem.*, 36 (1958) 59.
21 A. W. WITKOWSKI AND W. MOFFITT, *J. Chem. Phys.*, 33 (1960) 872.
22 E. G. McRAE, *Australian J. Chem.*, 14 (1961) 329, 344, 354; 16 (1963) 295, 315.
23 R. G. FULTON AND M. GOUTERMAN, *J. Chem. Phys.*, 35 (1961) 1059; 41 (1964) 2280.
24 M. GOUTERMAN, *J. Chem. Phys.*, 42 (1965) 351.
25 W. SIEBRAND, *J. Chem. Phys.*, 40 (1964) 2223.
26 W. GOAD, *J. Chem. Phys.*, 38 (1963) 1245.
27 A. S. DAVYDOV, *Zh. Eksperim. i Teor. Fiz.*, 18 (1948) 210.
28 J. PERRIN, *2me Conseil de chim. Solvay*, Gauthier-Villars, Paris, 1925, p. 322.
29 F. PERRIN, *Ann. Phys. (Paris)*, 17 (1932) 283.
30 Z. BAY AND R. M. PEARLSTEIN, *Proc. Natl. Acad. Sci. (U.S.)*, 50 (1963) 962.
31 TH. FÖRSTER, *Ann. Physik*, 2 (1948) 55.
32 M. TRLIFAJ, *Czech. J. Phys.*, 8 (1958) 510.
33 V. M. AGRANOVICH AND A. N. FAIDYSH, *Opt. i Spektroskopiya*, 1 (1956) 885.
34 A. PRÖBSTL AND H. C. WOLF, *Z. Naturforsch.*, 18a (1963) 822.
35 I. KETSKEMETY, *Z. Naturforsch.*, 17a (1962) 666; 21a (1965) 82.
36 M. Z. MAKSIMOV AND I. M. ROZMAN, *Opt. i Spektroskopiya*, 13 (1962) 90; *Opt. Spectry. (USSR)*, 13 (1962) 49.
37 TH. FÖRSTER, *Z. Naturforsch.*, 4a (1949) 321.
38 M. D. GALANIN, *Zh. Eksperim. i Teor. Fiz.*, 28 (1955) 485; *Soviet Phys. JETP*, 1 (1955) 317.
39 B. IA. SVESHNIKOV, *Dokl. Akad. Nauk SSSR*, 111 (1956) 78; 115 (1957) 274.
40 M. LEIBOWITZ, *J. Phys. Chem.*, 69 (1965) 1061.

41 V. V. ANTONOV-ROMANOVSKY AND M. D. GALANIN, *Opt. i Spektroskopiya*, 3 (1957) 389.
42 I. M. ROZMAN, *Opt. i Spektroskopiya*, 4 (1958) 536.
43 K. B. EISENTHAL AND A. S. SIEGEL, *J. Chem. Phys.*, 41 (1964) 652.
44 A. Y. KURSKII AND A. S. SELIVANENKO, *Opt. i Spektroskopiya*, 8 (1960) 643; *Opt. Spectry. (USSR)*, 8 (1960) 340.
45 S. F. KILIN, M. S. MIKHELASVILI AND I. M. ROZMAN, *Opt. i Spektroskopiya*, 16 (1964) 1063; *Opt. Spectry. (USSR)*, 16 (1964) 576.
46 J. FEITELSON, *J. Chem. Phys.*, 44 (1966) 1497.
47 M. INOKUTI AND F. HIRAYAMA, *J. Chem. Phys.*, 43 (1965) 1978.
48 E. RABINOWITCH AND L. F. EPSTEIN, *J. Am. Chem. Soc.*, 63 (1941) 69.
49 TH. FÖRSTER AND E. KÖNIG, *Z. Elektrochem.*, 64 (1957) 344.
50 J. LAVOREL, *J. Phys. Chem.*, 61 (1957) 1600; *J. Chim. Phys.*, 55 (1958) 905.
51 S. BASU, *J. Chim. Phys.*, 62 (1956) 827.
51a M. ASHRAF EL-BAYOUMI, *Dissertation*, Florida State University, Tallahassee, Florida, 1961.
52 G. SCHEIBE, *Angew. Chem.*, 49 (1936) 563; 50 (1937) 51, 212.
53 E. E. JELLEY, *Nature*, 138 (1936) 1009; 139 (1937) 631.
54 R. M. HOCHSTRASSER, *J. Chem. Phys.*, 40 (1964) 2559.
55 D. P. CRAIG, *J. Chem. Soc.*, (1955) 2302.
56 H. C. WOLF, *Z. Naturforsch.*, 13a (1958) 414.
57 M. S. BRODIN AND S. V. MARISOVA, *Opt. i Spektroskopiya*, 10 (1961) 473; *Opt. Spectry. (USSR)*, 10 (1961) 242.
58 D. FOX AND S. YATSIV, *Phys. Rev.*, 108 (1957) 938.
59 R. SILBEY, J. JORTNER AND ST. A. RICE, *J. Chem. Phys.*, 42 (1965) 1515.
60 G. CARIO AND J. FRANCK, *Z. Physik*, 11 (1922) 161.
61 J. PERRIN AND (MLLE.) CHOUCROUN, *Compt. Rend.*, 184 (1927) 1097.
62 TH. FÖRSTER, *Z. Elektrochem.*, 53 (1949) 93.
63 I. KETSKEMÉTY, *Acta Phys. Acad. Sci. Hung.*, 10 (1959) 429.
64 A. BUDO AND I. KETSKEMÉTY, *Acta Phys. Acad. Sci. Hung.*, 14 (1962) 167.
65 B. YA. SVESHNIKOV, P. I. KUDRYASHOV AND L. A. LIMAREVA, *Opt. i Spektroskopiya*, 9 (1960) 203; *Opt. Spectry. (USSR)*, 9 (1960) 107.
66 A. KAWSKI, *Acta Phys. Polon.*, 24 (1963) 641.
67 M. D. GALANIN, *Izv. Akad. Nauk SSSR, Ser. Fiz.*, 15 (1951) 543.
68 L. A. KUZNETSOVA AND B. YA. SVESHNIKOV, *Opt. i Spektroskopiya*, 4 (1957) 55.
69 E. J. BOWEN AND B. BROCKLEHURST, *Trans. Faraday Soc.*, 49 (1953) 1131; 51 (1955) 774.
70 E. J. BOWEN AND R. LIVINGSTON, *J. Am. Chem. Soc.*, 76 (1954) 6300.
71 W. R. WARE, *J. Am. Chem. Soc.*, 83 (1961) 4374.
72 A. G. TWEET, W. D. BELLAMY AND G. L. GAINES, *J. Chem. Phys.*, 41 (1964) 2068.
73 R. G. BENNETT, *J. Chem. Phys.*, 41 (1964) 3037.
74 O. SCHNEPP AND M. LEVY, *J. Am. Chem. Soc.*, 84 (1962) 172.
75 S. A. LATT, H. T. CHEUNG AND E. R. BLOUT, *J. Am. Chem. Soc.*, 87 (1965) 995.
76 K. H. DREXHAGE, M. M. ZWICK AND H. KUHN, *Ber. Bunsenges. phys. Chem.*, 67 (1963) 62.
77 H. KUHN *et al.*, in: *Optische Anregung organischer Systeme*, Verlag Chemie, Weinheim/Bergstr., 1966, p. 639.
78 R. G. BENNETT, *J. Chem. Phys.*, 41 (1964) 3048.
79 V. L. ERMOLAEV AND E. B. SVESHNIKOVA, *Izv. Akad. Nauk SSSR, Ser. Fiz.*, 26 (1962) 29.

80   V. L. ERMOLAEV AND E. B. SVESHNIKOVA, *Opt. i Spektroskopiya*, 14 (1964) 587; *Opt. Spectry. (USSR)*, 14 (1964) 320.
81   R. E. KELLOGG, *J. Chem. Phys.*, 41 (1964) 3046.
82   A. TERENIN AND V. L. ERMOLAEV, *Dokl. Akad. Nauk SSSR*, 85 (1952) 547; *Trans. Faraday Soc.*, 52 (1956) 1042.
83   V. L. ERMOLAEV, *Usp. Fiz. Nauk*, 80 (1963) 3; *Soviet Phys. Usp.*, 80 (1963) 333.
84   H. L. BÄCKSTRÖM AND K. SANDROS, *Acta Chem. Scand.*, 12 (1958) 823; 14 (1960) 48.
85   K. SANDROS AND H. L. J. BÄCKSTRÖM, *Acta Chem. Scand.*, 16 (1962) 958.
86   K. SANDROS, *Acta Chem. Scand.*, 18 (1964) 2335.
87   F. WILKINSON, *Discussions Faraday Soc.*, 27 (1959) 96.
88   G. PORTER AND F. WILKINSON, *Proc. Roy. Soc. (London)*, *Ser. A*, 264 (1961) 1.
89   J. B. FARMER, C. L. GARDNER AND C. A. MCDOWELL, *J. Chem. Phys.*, 34 (1961) 1058.
90   K. B. EISENTHAL AND R. MURASHIGE, *J. Chem. Phys.*, 39 (1963) 2108.
91   S. SIEGEL AND K. B. EISENTHAL, *J. Chem. Phys.*, 38 (1963) 2785.
92   R. W. BRANDON, R. E. GERKIN AND C. A. HUTCHISON JR., *J. Chem. Phys.*, 37 (1962) 447.
93   B. SMALLER, E. C. AVERY AND J. R. REMKO, *J. Chem. Phys.*, 43 (1965) 922.
94   S. SIEGEL AND H. JUDEIKIS, *J. Chem. Phys.*, 41 (1964) 648.
95   E. GAVIOLA AND P. PRINGSHEIM, *Z. Physik*, 24 (1924) 24.
96   F. PERRIN, *Compt. Rend.*, 192 (1931) 1727.
97   TH. FÖRSTER, *Fluoreszenz organischer Verbindungen*, Vandenhoek und Ruprecht, Göttingen, 1951, p. 244.
98   A. WINTERSTEIN AND K. SCHÖN, *Naturwiss.*, 22 (1934) 237.
99   E. I. BOWEN, *Nature*, 142 (1938) 108.
100 L. E. LYONS AND J. W. WHITE, *J. Chem. Phys.*, 29 (1958) 447.
101 K. W. BENZ AND H. C. WOLF, *Z. Naturforsch.*, 19a (1964) 177.
102 S. C. GANGULY AND N. K. CHAUDHURY, *Rev. Mod. Phys.*, 31 (1959) 990.
103 J. JORTNER, ST. A. RICE, J. L. KATZ AND S. I. CHOI, *J. Chem. Phys.*, 42 (1965) 309.
104 G. C. NIEMAN AND G. W. ROBINSON, *J. Chem. Phys.*, 39 (1963) 1298.
105 S. SINGH, W. J. JONES, W. SIEBRAND, B. P. STOICHEFF AND W. G. SCHNEIDER, *J. Chem. Phys.*, 42 (1965) 330.
106 P. AVAKIAN AND R. E. MERRIFIELD, *Phys. Rev. Letters*, 13 (1964) 541.
107 R. G. KEPLER AND A. C. SWITENDICK, *Phys. Rev. Letters*, 15 (1965) 56.
108 G. W. ROBINSON, *Ann. Rev. Phys. Chem.*, 15 (1964) 311.

Chapter III

Charge Transfer in Biology*

Section a

Donor–Acceptor Complexes in Solution

F. J. BULLOCK**

Laboratory of Chemical Biodynamics, University of California, Berkeley, Calif. (U.S.A.)

1. Introduction

In the context of biochemistry, the topic of charge transfer has been most closely associated with the donor–acceptor complexes. The research activity on the chemical and physical aspects of donor–acceptor complexes has been impressively on the increase, and, quite like a dose–response curve, interest in the biological implications of the new results has closely followed. We have attempted to present the topic of transfer of charge from a somewhat wider frame of reference than complex formation. Yet, clearly this topic, which can span the entire range of possibilities from weak interactions to full-fledged redox reactions in organic, inorganic or mixed systems, is so large that it could not be completely covered in less than a monograph. As a result, the decision as to what aspects would be emphasized largely reflects current interests, mostly our own. In Section *a* we are presenting solution properties of the donor–acceptor complexes. Two books and a review

* The preparation of this chapter was sponsored, in part, by the U.S. Atomic Energy Commission.
** Present address: Arthur D. Little Inc., Cambridge, Mass. (U.S.A.), NIH Postdoctoral fellow, 1964–65.

References p. 143

[81]

dealing with the general area of donor–acceptor complexes have recently appeared, covering the topic from the point of view of the physical chemist[1,2] and from the point of view of the organic chemist[3]. Another review[4] dealing mainly with reactions, has recently appeared. Despite the availability of these, it nevertheless seemed well for us to cover enough of the fundamental aspects of the topic to make this chapter more or less self-contained. However, particular emphasis has been placed on the more recent literature and those areas which seemed most germane to our purpose—a consideration of the biological aspects of the topic. Our literature survey was concluded in July, 1965. Szent-Gyorgyi's recent book[5] is also largely devoted to the possible role of donor–acceptor complexes in biology. Special aspects of charge-transfer interactions such as hydrogen bonding must here be totally neglected. Charge transfer in metal-ion complexes, a large topic in its own right, is treated only superficially.

The second part of the chapter (p. 149) will attempt to survey transfer of charge from the point of view of semiconduction and, in particular, photo-conduction. At present, the surge of interest in these fields stems more from the activities of the solid-state chemists and physicists than from those of the biochemist. The situation insofar as biochemistry is concerned is not really well-defined. Yet, as the important role of organized structures in living systems continues to emerge, it seems well that biochemists in general should have an awareness of the recent advances in these areas.

2. Energetics

As confusion on occasion has arisen, we should first define our terms. For complexes of the sort to be considered here, resonance contributions of the type drawn in Fig. 1 will play a role in defining the energetics. In accord with the distinction made by Mulliken and Person[6] the term "charge-transfer complex" will be applied to those complexes where the resonance interaction, R_N, makes the predominant contribution to the overall stability of the complex in the ground state (see Fig. 2). The term "donor–acceptor complex" will be applied to those complexes where the resonance contribution to its overall stability is less than that due to other factors such as dispersion forces, dipole–dipole interactions, etc. In order that the concept of a donor–acceptor and charge-transfer complex remain *experimentally* distinct from other possible types of complex, the donor–acceptor and charge-transfer complexes should exhibit a charge-transfer absorption band. The origin and some characteristics of this absorption will be discussed below. This is a more restrictive definition than that adopted in a number of biochemical papers. In our view, the value of the donor–acceptor concept is in its being subject to an unambiguous experimental test. That test, in this case, is the appearance

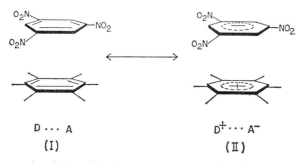

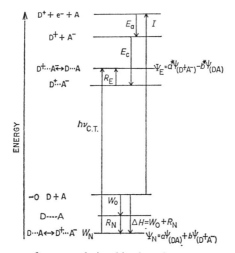

Fig. 1. Resonance description of the donor–acceptor complex of hexamethylbenzene and s-trinitrobenzene.

Fig. 2. Diagram of energy relationships in a donor–acceptor complex.

of a new absorption band and is, in fact, the only proper one. To those complexes experimentally demonstrable by any physical method, but not conforming to the above definitions by displaying a charge-transfer band, we apply the general term "molecular complex".

In Mulliken's valence-bond approach[7] the donor–acceptor complex, DA, is considered to be described by a wave function which may be written

$$\Psi = a\psi_{(DA)} + b\psi_{(D^+A^-)}$$

As in Fig. 1, we are stating in this relation that the ground state of the complex is described by taking some combination of the wave function $\psi_{(DA)}$ describing D $\cdots$ A (formula I, Fig. 1) and $\psi_{(D^+A^-)}$, describing D^+A^- (formula II, Fig. 1). In the term $\psi_{(DA)}$, the "no bond" wave function, we include all the dipole–dipole, Van der Waals–London and dipole–induced dipole forces. $\psi_{(D^+A)^-}$ is termed the "dative bond" wave function. The value of the coefficient a may be greater or less than b, but generally $a \gg b$. For the case $b \gg a$, to which we shall return later, the complex is ionic in the ground state, since D^+A^- predominates. In the case of a complex between a weak Lewis acid and a weak Lewis base, a third term $\psi_{(D^-A^+)}$ is added to the total ground-state wave function. In such cases $D^+A^- \leftrightharpoons D \cdots A \leftrightharpoons D^-A^+$ is a closer resonance description. The total wave function can then be taken as

$$\Psi = a\psi_{(DA)} + b\psi_{(D^+A^-)} + c\psi_{(D^-A^+)}$$

The energetics of charge-transfer complex formation are depicted in Fig. 2, but with neglect of possible solvation effects. The zero point of energy is taken as the energy of the separated donor and acceptor molecules. Bringing these molecules together to the normal equilibrium distance for the complex (about 3.5 Å) without permitting resonance interaction forces to operate results, in the case depicted, in a lowering of the energy by an amount, W_0. In practice this term could conceivably be a repulsive term. The resonance interaction with the charge-transfer structure D^+A^- further lowers the energy of the pair by an amount R_N. The total intermolecular binding energy, $\Delta H = W_0 + R_N$. Typical experimental ΔH values range from a few hundred calories to about 7 or 8 kcal·mole^{-1}. A breakdown of the contributions of W_0 and R_N to the total ΔH is given in Table I for a few cases. These experimental values are obtained by an indirect method. It involves an experimental determination of the dipole moment of the excited state of the complex and working through a lengthy series of equations which have been derived by Briegleb (see footnote a of Table I for reference).

By addition of a proper sized quantum of light (the charge-transfer energy $h\nu_{CT}$), a nearly complete transfer of an electron from donor to acceptor may be induced with formation of an excited state of greater polarity. An expression for the energy of the excited state is readily derived from Fig. 2, as

TABLE I

EXPERIMENTAL BREAKDOWN OF THE ENERGY OF FORMATION IN
SOME DONOR–ACCEPTOR COMPLEXES[a]

Acceptor						
Donor:	Hexamethyl-benzene	2,3,5,6-Tetra-methylquinol[b]	Hexamethyl-benzene	Durol	Hexamethyl-benzene	Durol
ΔH	−4.7	−4.0	−5.35	−4.4	−7.75	−5.5
W_0	−2.5	−2.65	−3.4	−2.5	−5.1	−2.8
R_N	−2.2	−1.35	−1.95	−1.9	−2.7	−2.7

[a] Energies are in kcal, data are from Ref. 1, p. 25, where experimental details and further
examples may be found.
[b] Durol.

follows: The ionization potential, I, of the donor corresponds to $D \to D^+ + e$,
and the electron affinity, E_a, to $e + A \to A^-$. The coulomb-energy term,
E_C, and the resonance interaction, R_E, contribute as indicated in the figure,
and the energy of the excited state becomes, by adding the individual
contributions

$$W_E = I_D - E_a - E_C + R_E$$

The reader may question why the resonance interaction raises the energy of
the excited state but lowers that of the ground state (cf. Fig. 3). A purely
mechanical explanation lies in the fact that the selection rule for the charge-
transfer transition requires the wave function of the excited state to be an
odd function (interchanging terms changes the sign of the function). The
wave function must, therefore, be taken as

$$\Psi = a\psi_{(D^+A^-)} - b\psi_{(DA)}$$

where the negative sign assures oddness. In the standard variational calculation of the energy by the Ritz method[8], the negative sign in this wave function
causes the second-order perturbation term (resonance-energy term) in the
expression for the energy to be of opposite sign from the resonance-energy
term of the ground state. However, real understanding comes by considering
electron-repulsion effects. For the ground state, the perturbation is a delocalization of a donor electron into an empty acceptor orbital. The decrease
in electron repulsion lowers the energy. For the ionic excited state, the per-

turbation is the introduction of a little ground-state localization. The electron repulsion increases, and the energy is raised.

The energy of the charge-transfer transition is then

$$W_E - W_N = h\nu_{CT} = I_D - E_a - E_C + R_E - W_0 - R_N = I_D - E_a + \Delta$$

where several factors have been absorbed in the term Δ, which is empirically observed to be approximately constant for many complexes having a common donor or acceptor.

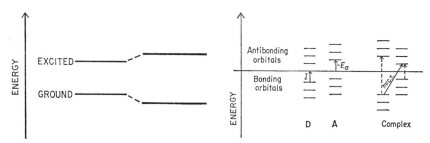

Fig. 3. Effect of resonance interaction on the energies of the ground and excited states.

Fig. 4. Schematic diagram of energy relationships in a donor–acceptor pair according to the molecular-orbital description (perturbation effects exaggerated).

In the molecular-orbital approach to the description of donor–acceptor complexes favored by Dewar[9], the orbitals are considered to extend over the entire complex. For weak complexes, the donor–acceptor interaction between the highest filled bonding orbitals of donor and the lowest empty orbitals of acceptor is considered as a perturbation. The new orbitals for the weak complexes will be very similar to the orbitals of donor and acceptor separately, but with the energy levels of donor slightly lowered and those of acceptor slightly raised (Fig. 4). In molecular-orbital theory the ionization potential of the donor is approximately equal to minus the energy of the highest occupied MO and the electron affinity of acceptor to the energy of the lowest unoccupied MO. The energy of the first charge-transfer transition (solid arrow in Fig. 4) is given by

$$\Delta W = h\nu_{CT} = I_D - E_a + \Delta$$

where Δ is a term resulting from the perturbation. As it should be, this expression is identical to that obtained in the valence-bond approach outlined above. The dashed arrows in Fig. 4 represent the spectroscopic transitions characteristic of the donor and the acceptor. It is important to realize the following points. Perturbations resulting from neighbor interactions can occur in complexes in the absence of a resonance interaction such as in Fig. 1 and can cause small shifts in the absorption maxima of the components. Hypochromic effects and band broadening can be explained without invoking an intermolecular donor–acceptor interaction[10].

In the MO description the charge-transfer interaction will be inversely proportional to the difference in energy between interacting orbitals. Extensive tabulations of the results of Hückel calculations for many molecules of interest in biochemistry have been published in the Pullmans' recent book[11], and may serve as a guide for the experimentalist. But, if need be, one should not hesitate to resort to chemical considerations in predicting donor–acceptor compounds. Clearly, species capable of undergoing a facile oxidation (a measure of ease of electron loss) such as iodide ion, hydroquinones, amines and phenols should be donors. The donor strength of π-systems should be increased by electron-donating groups such as alkyl, $-NR_2$, $-OH$, and $-O-CH_3$. Species capable of undergoing a ready chemical reduction (electron gain) such as iodine, oxygen, quinones and riboflavin should be acceptors. Acceptor strength should be increased by electron-withdrawing substituents such as $-CF_3$, $-CN$, $-SO_3H$ and $-COOR$.

Using a given acceptor, the linear correlations anticipated above between the energy of the charge-transfer transition and the ionization potential or

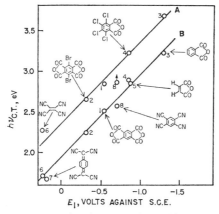

Fig. 5. Dependence of the energy ($h\nu_{CT}$) on the first half-wave reduction potential (in acetonitrile) of acceptor for complexes with A, hexamethylbenzene and B, pyrene. (From ref. 14)

calculated energy of the highest occupied molecular orbital of donor have been observed[12,13]. Experimental values for electron affinities are accessible with difficulty. However, the expected relationships between the energy of the charge transition and first half-wave reduction potential[14] (Fig. 5) and/or the calculated energies of the lowest unoccupied molecular orbital of the acceptor[16] have been observed.

An important concern in dealing with complexes is the matter of stability or trends in stabilities. The free energy of complex formation is not intimately related to such parameters as ionization potential, electron affinity, or charge-transfer transition energy. The reasons are several. First, a considerable contribution to the enthalpy part of the free-energy term is from Van der Waals or other forces (*cf.* Fig. 2 and Table I). It is not expected that these forces will be related in any simple way to the above parameters. The entropy contribution to the free energy probably strongly reflects solvation differences and again no simple correlations are anticipated. The scatter of points in Figs. 6, 7 and 8 undoubtedly reflects such effects as we have described. Never-

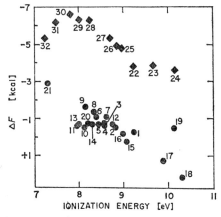

Fig. 6. Free energy of formation, ΔF, in relation to the ionization potential of donor for donor–acceptor complexes of iodine. ● = aromatic and aliphatic hydrocarbon donors; ◆ = amine donors. (Data from ref. 1)

1, benzene	9, hexamethylbenzene	17, cyclohexene	25, CH_3NH_2
2, toluene	10, naphthalene	18, 2,3-dimethylbutane	26, $C_2H_5NH_2$
3, *o*-xylene	11, 1-methylnaphthalene	19, 1-bromobutane	27, $nC_4H_9NH_2$
4, *m*-xylene	12, styrene	20, anisole	28, $(CH_3)_2NH$
5, *p*-xylene	13, stilbene	21, *N,N*-dimethylaniline	29, $(C_2H_5)_2NH$
6, mesitylene	14, biphenyl	22, pyridine	30, $(CH_3)_3N$
7, durol	15, chlorobenzene	23, α-picoline	31, $(C_2H_5)_3N$
8, pentamethylbenzene	16, bromobenzene	24, NH_3	32, $(nC_3H_7)_3N$

theless, the rough correlations evident from the figures between ionization potential, electron affinity, or charge-transfer energy, and complex stability may occasionally be useful for making qualitative predictions.

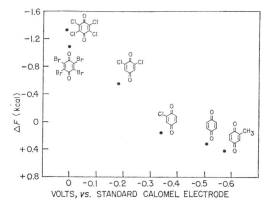

Fig. 7. Free energy of formation, ΔF, in relation to first half-wave potential of acceptor in donor–acceptor complexes with hexamethylbenzene as donor. Thermodynamic data from ref. 15; polarographic data from ref. 17.

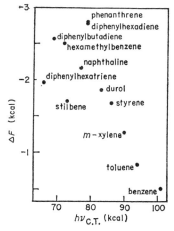

Fig. 8. Free energy of formation, ΔF, in relation to charge-transfer energy $h\nu_{CT}$, of a series of complexes with trinitrobenzene. (Data from ref. 1)

3. Characteristics of the charge-transfer absorption band

The charge-transfer band is always broad, frequently asymmetric, and devoid of vibrational fine structure, even at low temperatures. The broadness and lack of fine structure are undoubtedly due to small variations in the equilibrium distance between partners. Because of the broad and asymmetric character of the charge-transfer absorption, it may often be difficult to determine the position of the absorption maximum precisely. However, fine structure has on at least one occasion been observed in the charge-transfer band. This is in the donor–acceptor complexes of menadione with 3,5-diiodotyrosine or 3,5-dibromotyrosine[18] (Fig. 9). This was attributed, in

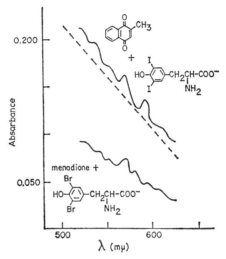

Fig. 9. Charge-transfer band of complexes of 3,5-diiodotyrosine and 3,5-dibromotyrosine with menadione. Use of phenolate and menadione alone gave no absorption in this region. Dotted line indicates anticipated featureless band. (From ref. 18)

this atypical case, to a superposition of the single–triplet transition of the quinone which becomes increasingly allowed due to the fact that the halogen atoms cause a breakdown of the spin-selection rules by spin–orbit coupling. The evidence in favor of assigning a very weak absorption at 535 mμ in p-benzoquinone to a singlet–triplet transition has recently been reviewed[19].

Briegleb has noted[20] that a fairly good empirical correlation exists between v_{max} (cm^{-1}) and the position of half-maximum height of the charge-transfer absorption band, $v_{max} - v_{\frac{1}{2}} \approx 0.104\ v_{max}$. For the few cases of complexes of biological molecules where sufficient data have been reported to enable comparison within a series, we find that the band shapes deviate from Briegleb's correlation. Whether this is attributable to the fact that the charge-transfer bands, frequently reported as difference spectra, are not authentic or due to other reasons is not clear. Briegleb also reports[20] that within a series, the half-width at half-height varies inversely with the heat of formation of the complex. Thus, weak complexes will show the broadest charge-transfer absorption bands.

Some donor–acceptor complexes show two distinct maxima in the charge-transfer band. Examples are the complexes of certain substituted benzenes and tetracyanoethylene (TCNE)[21]. The most likely explanation of the multiple maxima is as follows: In molecules of high symmetry such as benzene the highest occupied molecular orbitals are degenerate (Fig. 10). On sub-

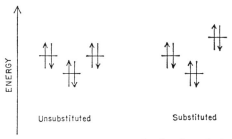

Fig. 10. Schematic representation of the energy levels of a substituted benzene and an unsubstituted benzene.

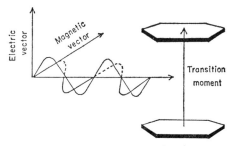

Fig. 11. Schematic diagram of incident light with electric vector polarized parallel to charge-transfer transition moment.

stitution of the benzene ring, the degeneracy is removed, and the two orbitals now split to an extent depending on the number of substituents, their nature and relative position on the ring. The resulting two electronic levels have two different ionization energies, and the transitions from each are no longer equienergetic.

In crystals, where the orientation of the molecules of the donor–acceptor complex relative to the crystal axis is fixed, it is possible to determine the direction of polarization of the transition moment of the charge-transfer absorption by using polarized light. The light absorption by the crystal is greatest when the electric vector of the incident polarized light is parallel to the direction of polarization of the transition moment of the absorption band (Fig. 11). Historically, such a study of the quinhydrone (p-benzoquinone–quinol) complex[22] provided important support for Mulliken's theory which predicted that the direction of charge transfer should be from donor to acceptor ring. Recently a reinvestigation of this complex using a technique for obtaining polarized reflection spectra[23] rather than polarized absorption

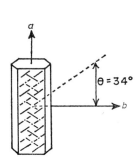

Fig. 12. Morphology and molecular orientation in a quinhydrone crystal. —·—·—, benzoquinone; ———, hydroquinone.

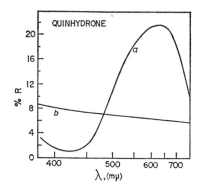

Fig. 13. Reflection spectra obtained on the prominent (001) face of the quinhydrone crystal. (a) light polarized parallel to axis a; (b) light polarized parallel to axis b. (From ref. 23)

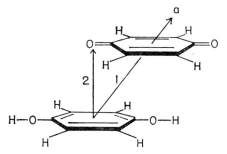

Fig. 14. Direction of polarization of the charge-transfer transition in quinhydrone. New assignment is arrow number one.

spectra has shown the old data to be somewhat in error. In the quinhydrone crystal it is known from X-ray studies that the molecules are oriented with their planes parallel, and tilted 34° from the direction of the needle axis of the crystal (Fig. 12). The reflection spectra obtained from the prominent face of the crystal and the direction of the polarization of the light are reproduced in Fig. 13. From the essentially plane curve observed for light polarized along the *b* axis it was concluded that the charge-transfer moment was exclusively along the needle (*a*) axis and therefore directed between the centers of the six-membered ring (Fig. 14). Nakamoto's study[22] had indicated a small absorption component along the *b* axis and a direction of polarization of the transition moment perpendicular to the plane of the rings (arrow 2).

A result completely analogous to that obtained by Anex and Parkhurst[23] has been obtained for a complex of coronene and chloranil by Ilten and Sauer[23a] in this laboratory.

4. Paramagnetic complexes and electron-transfer reactions

Complete transfer of an electron from donor to acceptor can occur thermally with formation of an ionic species. This corresponds to the $b \gg a$ case referred to on p. 84. A potential-energy diagram illustrating this situation is given in Fig. 15. In solution, the thermally induced electron transfer may go through a "tight" ion pair on the way to solvent separated ions.

$$[DA \cdots D^+A^-]_{solvent} \rightleftharpoons [D^+A^-]_{solvent} \rightleftharpoons [D^+]_{solvent} + [A^-]_{solvent}$$

The radical ions formed should show an ESR (electron-spin resonance) signal, but frequently a signal characteristic of only one species is observed.

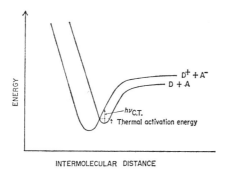

Fig. 15. Potential-energy diagram illustrating possible thermal and photo-induced electron transfer from donor to acceptor.

This is presumably due to disproportionation of one radical. A further point of interest regarding these electron-transfer reactions is that the activation energy for the thermal reaction may be lower than the photochemical energy needed for electron transfer (as in Fig. 15). The photochemical pathway can lead to the ions in higher vibrational levels.

One of the most studied examples of complete electron transfer in a complex where both donor and acceptor are organic species is the tetramethyl-*p*-phenylenediamine–chloranil case. In polar solvents (see also the section on solvent effects, p. 101) the absorption spectrum is a superposition of what one observes separately for the Wurster's blue cation radical (III) and chloranil radical anion (IV). A recent report contains a survey of the older literature concerning electron-spin resonance studies of this complex[24]. Often the free-radical species initially produced in such reactions undergo further slow chemical changes, not infrequently to intractable products.

Crystal violet

Reaction between *N,N*-dimethylaniline and chloranil in polar solvents yields a violet material with an absorption spectrum corresponding to that of the crystal violet cation[25]. The diamagnetic complex (D $\cdots$ A) and semiquinone radical were identified as intermediates. The overall course of the reaction is apparently as given above. The intermediate steps must involve an intermolecular methyl migration.

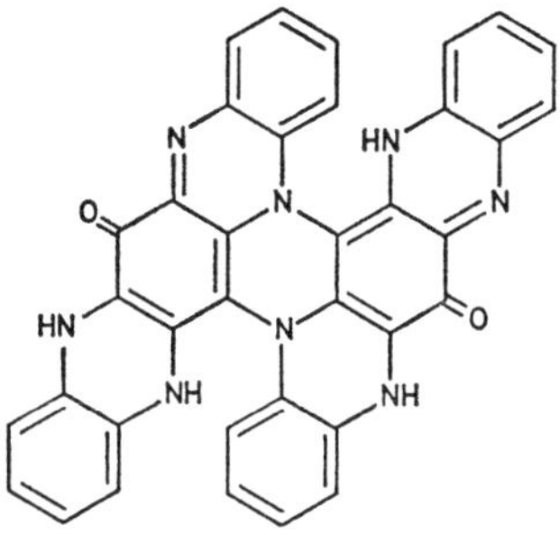

(V)

It is of interest that unsubstituted *o*-phenylenediamine and chloranil react[26] under nitrogen with displacement of halogen to yield a dimeric quinoneamine of suggested structure (V) (four moles of amine, two of quinone). Therefore, in dealing with chloranil as an acceptor, one must also

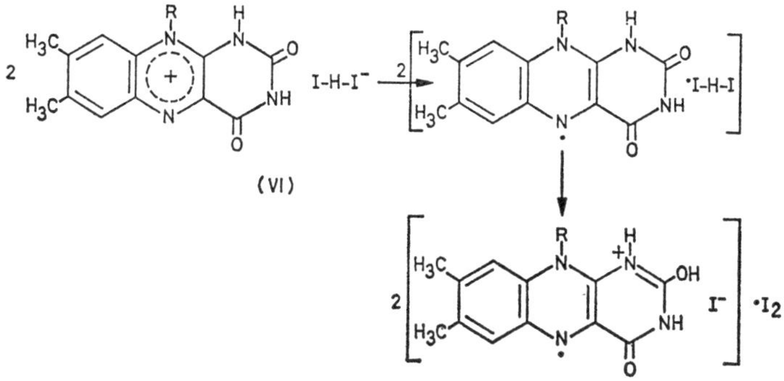

(VI)

be cognizant of possible displacement reactions, particularly with amine donors. The displacement may occur subsequent to complex formation.

Recently a remarkable complex of riboflavin hydroiodide and HI has been reported by Fleishman and Tollin[27]. This material, obtained by evaporating a solution of riboflavin in 47% hydroiodic acid–ethanol mixtures occurs as pink platelets. The analytical data suggest the stoichiometric composition, riboflavin hydroiodide : HI (1 : 1). The visible absorption spectra of the solution and solid correspond to that of protonated riboflavin semiquinone. The material shows an ESR signal characteristic of the semiquinone and of such intensity that the material must be 100 % radical. It seems quite likely that this material is derived from the hydrogen diiodide salt of riboflavin (VI) by an electron transfer from the anion. This complex is then similar to the complex of *p*-phenylenediamine and chloranil in this respect. Hydrogen dibromide and dichloride salts of ammonium ions are known[28] and a tetra-alkylammonium hydrogen diiodide (VII) has recently been prepared[29]. Hydrogen dibromide salts of carbonium ions (*e.g.* VIII) were also reported[29] and show a charge-transfer absorption band.

$$(nC_4H_9)_4 \overset{+}{N} \ I\text{-}H\text{-}I^-$$

(VII)

(VIII)

Some other oxidation reactions may follow the general course of complex formation followed by electron transfer. The oxidation of duroquinol by 7,7,8,8-tetracyanoquinondimethide (TCNQ), an excellent electron acceptor, in the presence of a proton acceptor has been reported[30].

Duroquinol   TCNQ

As with chloranil, one should be aware of the possibility of a displacement reaction with TCNQ or TCNE (tetracyanoethylene). Indole has been observed to react with TCNE[31] in solution at ordinary temperatures. A reaction with *N*-alkylanilines is also observed[32]. Amines have been observed to displace a cyano group from TCNE[33] and TCNQ[34]. Formation of these latter products is accompanied by a pronounced bathochromic shift in the absorption spectrum which should not be interpreted as due to complex formation.

Mercaptide is oxidized anaerobically to disulfide in the presence of nitrobenzene[35] or other good electron acceptor[36]. The nitrobenzene anion radical may be detected by ESR.

Russell and his coworkers[37] have noted a redox reaction between the monoanion of 1,4-quinol and nitrobenzene. Both product species were detected by

ESR. They were also able to detect the nitrobenzene anion radical when carban-ions were used as donors.

Charge transfer has also been used to initiate polymerizations. N-Vinyl-carbazole (IX) is polymerized to poly(N-vinylcarbazole) in the presence of p-chloranil, tetracyanoethylene and tetracyanoquinondimethide, among others[38]. There is some evidence that complex formation precedes electron

transfer as suggested above for the reaction leading to crystalviolet. A mechanism such as that outlined is favored by the authors. It appears, however, that electron delocalization may be adequate to induce polymerization. Trinitrostyrene (X) copolymerizes exothermically with vinylpyridines (XI) on mixing solutions of the two[39]. Free-radical formation as an initiating step appears less likely here.

We cannot undertake here a systematic treatment of the many cases of electron transfer from zero valent metals to hydrocarbon acceptors capable of yielding anion radicals, although these processes involve transfer of charge in a broad sense[40]. Also beyond the somewhat arbitrary scope of this chapter

is a detailed consideration of the mechanisms of electron transfer between inorganic metal ions *via* inorganic or organic bridging groups[41].
For discussion of magnetic resonance methods for determining the rates of electron exchange between radicals of the type we have considered in this section, one may consult, among others, the recent book of Caldin[42] or the chapter of Fraenkel[43].

5. Equilibrium constants

It is not our purpose to review in detail here the numerous prodecures available for determining equilibrium constants, as they have been amply treated elsewhere[44,45]. As a number of authors have recently emphasized, the determination of an equilibrium constant and a molar extinction coefficient should be part of the characterization of any donor–acceptor complex. This is not an entirely trivial matter, however, and we will discuss a few recent papers in which some of the pitfalls have been exposed.

Person[46] has discussed the use of the Scott equation (1) in some detail.

$$\frac{[D]\,[A]l}{\Delta A_k} = \frac{1}{K\varepsilon} + \frac{1}{\varepsilon}\,[D] \tag{1}$$

ΔA_k is the absorption due to complex at wavelength k, K is the equilibrium constant, ε the molar extinction of complex, l the optical pathlength and the remaining terms represent donor and acceptor concentrations. The absorption due only to complex (ΔA), which is the absorption of the mixture at wavelength k minus the sum of the absorption due to components, is determined for a series of donor concentrations. The left side of equation (1) is then plotted *versus* donor concentrations and extrapolated to the intercept to determine K. For weak complexes, since the concentration of complex is small, the points plotted as the left side of equation (1) may not be very different from the intercept. Difficulty is then experienced in obtaining a reliable non-zero initial slope in the plot of the data. This slope is obviously important since the determination of K involves an extrapolation. Person[46] has set the optimal concentration range for use of equation (1) (for the case of excess donor) at

$$0.1\left(\frac{1}{K}\right) < [D] < 9\left(\frac{1}{K}\right)$$

where K is the equilibrium constant. These limits may be a little flexible, but one should proceed by first obtaining a rough value for K, then planning optimum concentration ranges for final runs.

A different approach to a systematic study of the effects of experimental errors has also appeared[47]. A computer program was developed which enabled ready calculation of the equilibrium constant from input experimental data. Beginning with synthetic data (no errors) small errors were deliberately introduced into the input data—for example, amounting to a weighing error of 0.3 mg in 20–500 mg, or an instrumental error of ± 0.003 absorbance units. Recalculation of the constant revealed that for certain concentration situations the determined equilibrium constant could be extremely sensitive to small experimental errors. The same conclusion was reached when K was determined by a graphical method with the same data. This again emphasizes the need for careful planning of experimental conditions.

In a second study[48], again employing synthetic data, multiple equilibria were considered. The results indicate that the presence of some 1 : 2 complex may not be detected as a deviation of the data from a linear plot of the usual sort. Johnson and Bowen specifically considered the data plots obtained with eqn. (2) (the Benesi–Hildebrand equation), but the conclusions will presumably apply to all similar equations.

$$\frac{[A]}{\varDelta A} = \frac{1}{\varepsilon} + \frac{1}{K\varepsilon[D]} \tag{2}$$

Experimentally, the presence of higher order complexes may be manifest as variations of the determined equilibrium constant with wavelength, or as a variation of the integrated intensity of the charge-transfer band with temperature. Practically speaking, the effects are often small and one could probably conclude that the *major* absorbing species is a 1 : 1 complex if the data fit a linear plot of the Benesi–Hildebrand type. Methods are available for an independent experimental determination of the stoichiometry of a complex (*cf.* Refs. 44, 45).

6. Contact charge transfer

Mulliken's theory of charge-transfer spectra predicts that for weak complexes the molar extinction coefficient of the charge-transfer band should be low. There have been many difficulties in demonstrating this expectation experimentally. For weak complexes, attempts to determine both the equilibrium constant and extinction coefficient experimentally (for example, using Eqn. 1 or 2) often gave the point of intercept in the extrapolation as zero. This indicated either that as $K \to 0$, $\varepsilon \to \infty$, or *vice versa*, which was contrary to that anticipated from the theory. In an attempt to explain this difficulty Orgel and Mulliken[49] put forward the concept of contact charge transfer. In essence this theory stated that a charge-transfer transition could occur between two molecules [D,A], which just happen to be together through chance collisions without forming a real complex. The observed extinction coefficient would then be the sum of the actual extinction and some contribution from chance contacts. It has now been shown[50] that a satisfactory theory for weak complexes can be based on the idea of competition between solvation and complexing by writing the equilibrium expression as

$$DS_n + AS_m \rightleftarrows DAS_p + q\,S$$

where S represents the solvent. By taking proper account of solvent, it was shown that the relationship between ε and K anticipated from Mulliken's early theory could be observed. The upshot is that it now seems unnecessary to retain the concept of contact charge transfer.

7. Solvent effects

The effects of solvent on charge-transfer maxima have been reviewed by Murrell[51] and Reichardt[52], the latter in connection with empirical measures of solvent polarity. The rules of thumb are as follows: For complexes of the non-ionic type, the charge-transfer maximum should be red shifted in polar solvents, blue shifted in non-polar solvents. For ionic complexes, the opposite behavior is anticipated. However, a recent attempt[53] to correlate absorption maxima of complexes with solvent polarity has failed, and it appears there may be exceptions to the general rules.

Complex stability is also affected by solvent. Ionic complexes show a lower degree of association in polar solvents as Kosower[54] has observed for a series of pyridinium iodides. For complexes where little charge is transferred in the ground state, it is difficult to make a general statement concerning trends in stability. The situation varies from case to case, depending on the relative changes in solvation energies of components and complex. In cases where hydrogen bonding can play a role, the situation is further complicated.

In certain cases—for example, the tetramethyl-*p*-phenylenediamine complex with chloranil—one may obtain an ionic or non-ionic complex by varying solvent polarity[25].

8. Other useful physical methods for the study of complex formation

Many physical methods are useful for studying complexes. We emphasize, however, that not all these methods are adequate in characterizing a complex as of the donor–acceptor type.

(a) Infrared spectroscopy

Shifts in bond-stretching frequencies as a result of changes in bond strength can be anticipated in strong donor–acceptor complexes. Intensities are generally decreased, and in certain cases vibrations which are ordinarily symmetry forbidden may appear. For weak complexes, the spectra hardly differ from a superposition of the spectra of the components, a difficulty common to all the spectroscopic tools. Many specific cases have been reviewed in refs. 1 and 3. We will note only the recently described[55] complex of phenothiazine and chloranil (XII) as it is illustrative of a common result. In a Nujol mull or KBr pellet the quinone carbonyl peak appears at 6.4 μ (1560 cm^{-1}), shifted up from its usual 6.0 μ (1660 cm^{-1}) position. This immediately identifies the complex as of the ionic type.

(XII)

For complexes of donors with low ionization potential or acceptors of high electron affinity (or both) the charge-transfer band may be found in the near infrared. For example, the p-phenylenediamine–chloranil complex[25] shows a new absorption at 942 mμ in acetonitrile. In polar solvents the complex of β-carotene and iodine shows a new absorption at 1000 mμ. This complex is a 1:2 complex, characterized as $[(C_{40}H_{56})I^+]I_3^-$. The charge-transfer absorption is attributed to the moiety $(C_{40}H_{56} \rightarrow I^+)$ in which the I^+ is acting as a very powerful electron acceptor[56]. A band at 900 mμ has been assigned to a donor–acceptor complex of riboflavin and dihydroflavin[57,58].

(b) Nuclear magnetic resonance spectroscopy

Thus far only proton resonance appears to have been employed for the study of donor–acceptor complexes. Attempts to determine equilibrium constants with NMR have met with varying degrees of success. A derivation similar to that of the Benesi–Hildebrand Eqn. (2) readily leads to an equation suitable for use with NMR. Eqn. 3 is the form for donor in excess, the reciprocal of changes in observed chemical shifts of acceptor being plotted *versus*

$$\frac{1}{\Delta \text{obsd}_A} = \frac{1}{K\Delta_{DA}^A \,[D]} + \frac{1}{\Delta_{DA}^A} \tag{3}$$

the reciprocal of donor concentration. The term Δ_{DA}^A is the chemical shift of pure complex and corresponds to the molar extinction coefficient in the usual form of this equation. Hanna and Ashbaugh[59] have studied a number of complexes of aromatic donors with TCNQ. They find equilibrium constants in agreement with those obtained by other methods when donor concentrations are expressed in terms of molality. The reported failure of the method when concentrations are expressed as mole fractions is disconcerting. An attempt to determine with NMR the equilibrium constants for complexes of silver ions with a number of olefins has also failed[60].

A successful attempt to obtain an equilibrium constant by NMR is reported for a complex of iodine and phenylmethylsulfide[61]. In addition, the changes in chemical shift with temperature were sufficiently large in this case to enable determination of the thermodynamic parameters for complex formation. By using the relaxation times of the signals, Larsen and Allred deduce a lifetime of about 10^{-4} sec for the complex, surprisingly long for a complex in solution.

Magnetic resonance should prove a useful tool in facilitating identification of coordination sites in complexes. For example[62], in the aniline–iodine complex, a large selective downfield shift is observed for the N–H protons on complex formation. Such a large deshielding of these protons can only indicate that charge is indeed transferred from nitrogen and that this is the site of coordination with iodine.

(c) Polarography

It has long been appreciated that complex formation can affect the observed reduction potential of molecules[63]. However, little systematic study seems to have been reported on these effects for donor–acceptor complexes in general. Recently a polarographic study of several complexes, known to be of the donor–acceptor type, has been carried out in non-aqueous solvents. An attempt was made to employ shifts in half-wave potentials for the determination of equilibrium constants[64]. The method is analogous to that commonly used for metal ion complexes[65].

For a 1 : 1 complex, the situation in the polarographic experiment is described by Eqn. 4.

$$DA \underset{K_b}{\overset{K_f}{\rightleftharpoons}} D + A \overset{e^-}{\rightleftharpoons} A^- \qquad (4)$$

Provided A^-, the product of the polarographic reduction, has a negligible ability to function as an acceptor with D, a study of the shift in half-wave reduction potential of the $A + e^- \rightarrow A^-$ system with donor concentration may be related to the free energy of formation of DA. The conditions which must be satisfied for successful use of the method are (i) chemical equilibrium for the formation of DA must be rapidly achieved; (ii) the rate of electron transfer to A or (DA) at the electrode must be high; and (iii) the solvation energy of A^- must be unchanged by addition of the donor. The fulfillment of these conditions may be tested experimentally[65,66]. When these conditions are fulfilled an expression of the form (5) may be derived. Here we have written activity coefficients equal to one for

$$F_0 = \text{antilog}_{10} \left[\frac{0.4343F}{RT} \Delta E_{\frac{1}{2}} + \log_{10} \frac{I_s}{I_c} \right] = 1 + K_1[D] + K_2[D]^2 + \dots \qquad (5)$$

simplicity. F is the Faraday, I_s and I_c are the limiting currents for uncomplexed and complexed species, respectively, and K_1, K_2, etc. are the equilibrium constants for 1 : 1 complex formation, 1 : 2 complex formation, etc. All terms on the left are obtained from polarographic data and a plot of $(F_0 - 1)/D$ versus [D] yields a value for K_1 from the intercept (Fig. 16). A zero slope ($K_2 = 0$) is indicative of a 1 : 1 complex, the non-zero slope indicates 2 : 1 interactions. The constant K_2 may be obtained from this slope. Good agreement was observed between the constants obtained by this method and those obtained by the usual spectroscopic method for 1 : 1 complexes. It should be noted that the polarographic data indicated some 2 : 1 complex in certain instances where 1 : 1 complexes has been assumed in optical studies.

This is especially interesting in view of the results obtained by Johnson and Bowen[48] discussed previously.

The polarographic technique gives no information on the nature of forces leading to complex formation.

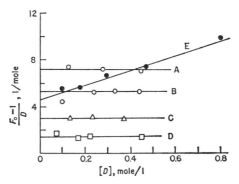

Fig. 16. Polarographic data for complex formation as a function of donor concentration (in CH_2Cl_2 at 25°). Hexamethylbenzene complexes of: A, tetracyanoethylene; B, tetracyanobenzene. Pyrene complexes of: C, chloranil; D, tetracyanobenzene; E, tetracyanoquinondimethide. (Data from ref. 64)

(d) Fluorescence and phosphorescence

Fluorescence techniques in particular are admirably suited for exploitation in biochemical research[67-69]. Detailed kinetic information as well as equilibrium constants may be obtained. It seems worth emphasizing, however, that while demonstration of fluorescence quenching does indicate an interaction, this does not itself demonstrate an interaction of the donor–acceptor type.

The energy absorbed in the charge-transfer transition may be emitted as a fluorescence. The spectrum of the fluorescence is quite generally a mirror image of the charge-transfer absorption spectrum. In energy-transfer studies the charge-transfer emission may be of considerable interest, but for diagnostic purposes the absorption spectrum is infinitely superior.

Phosphorescence from donor–acceptor complexes at low temperature is known, but the emission is invariably characteristic of the donor. Christodouleos and McGlynn[70] have reported a particularly interesting study. They observed a marked increase in the ratio of phosphorescence to fluorescence quantum yields with increasing concentration of acceptor in several complexes. The effects are particularly marked for polyhalogenated acceptors.

This again demonstrates an enhanced intersystem crossing from singlet to triplet levels in donor–acceptor complexes of halogenated materials due to spin–orbit coupling.

(e) Temperature-jump relaxation and flash techniques

The rate constants K_{12} and K_{21} in an equilibrium can often be determined by the temperature-jump relaxation method[71]. The method involves altering

$$D + A \underset{K_{21}}{\overset{K_{12}}{\rightleftarrows}} D \cdots A$$

suddenly the temperature of a system in equilibrium (2–10 degree jumps in 10^{-6} sec are common) and following the course of recovery to a new equilibrium by some rapid-detection method. It is a general method for the study of complexes. A complex of serotonin creatinine sulfate and riboflavin has been studied by this method[72]. The constant K_{12} is greater than $8 \cdot 10^{7}$ mole^{-1} sec^{-1}; K_{21} is greater than $2 \cdot 10^{5}$ sec^{-1}. Accordingly, the life-time of the complex in solution must be less than $5 \cdot 10^{-6}$ sec.

Flash spectroscopy has turned up the existence of donor–acceptor complexes of iodine atoms (produced by the flash) and benzenoid hydrocarbons (XIII)[73]. The existence of some sigma complex in these systems can probably

(XIII)   Sigma complex

not be ruled out with absolute confidence. For some of the hydrocarbons, complex formation was completely reversible, for others iodination occurred, implicating a sigma complex. In view of chemical similarities between the iodine atom and the mercaptyl radical RS· (e.g., both dimerize to a stable molecule, can be reduced to a stable anion, can be oxidized in the presence of water to an acid, are highly polarizable) one might anticipate similar complexes of the mercaptyl radical and biologically more interesting π systems. None has been identified thus far.

9. Effect of complex formation on reaction rates

From the previous discussions it might well be anticipated that the chemical properties of a donor or acceptor may be affected by the donor–acceptor interaction. Indeed, a number of cases are known where specific catalysis due to complexing with a donor or acceptor is observed. In the following we give some recent examples.

Complex formation may be expected to produce a rate enhancement in a reaction in which the free energy of the transition state is lowered more than that of the ground state by the interaction. The solvolysis of dimethyl benzylsulfonium chloride (XIV) has been observed to proceed three times faster with phenoxide than with hydroxide, although the latter is a much stronger base[74].

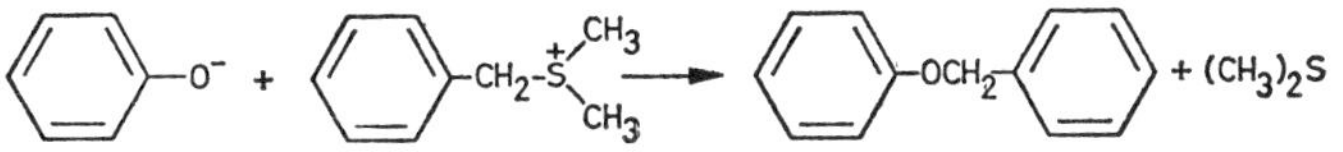

This rate enhancement appears to be due to π complexing in the transition state, such as in Fig. 17. In the solvolysis of trimethylsulfonium chloride, where π complexing is impossible, the rate is higher with hydroxide, as expected. This fact lends support to the complexing interpretation.

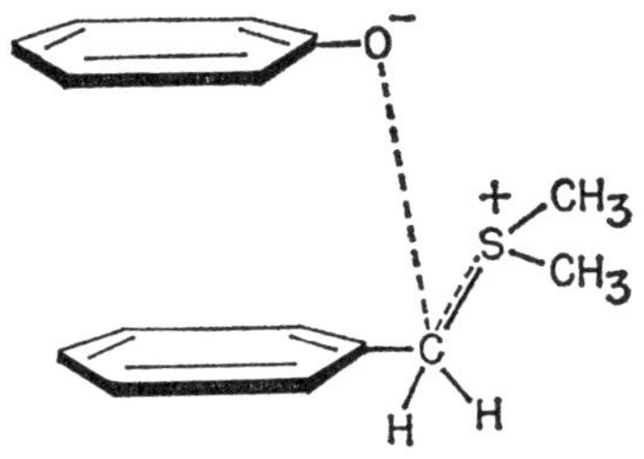

Fig. 17. Possible π complexing in the transition state during the solvolysis of benzyl-dimethylsulfonium ion by phenoxide.

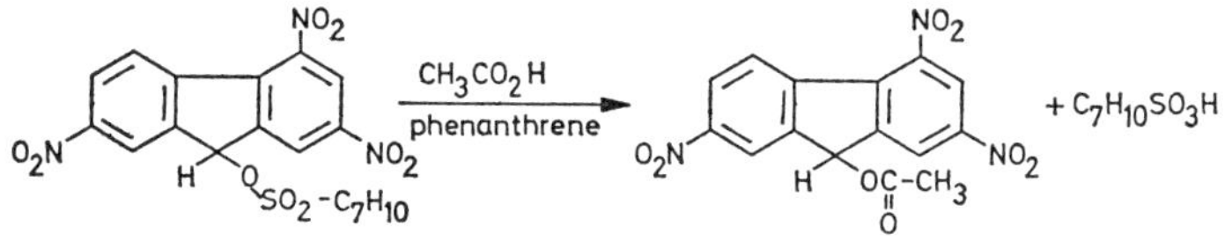

A detailed study of the acetolysis of 2,4,7-trinitro-9-fluorenyl-*p*-toluene-sulfonate (XV) has been carried out in the presence of phenanthrene[75,76]. The 2,4,7-trinitrofluorenyl ring system is a good acceptor and was chosen in an attempt to maximize possible catalytic effects. The kinetic data is successfully interpreted in terms of the simplest mechanisms involving $1:1$ complex formation (below). The ratio of the rate constants, $K_{complexed}/K_{uncomplexed}$, was found to be 21, indicating a very effective catalysis. The equilibrium

$$ROSO_2C_7H_{10}+Donor \xrightleftharpoons{K_T} ROSO_2C_7H_{10} \cdots D$$

$$HOAc \downarrow K_{uncomplexed} \qquad HOAc \downarrow K_{complexed}$$

$$ROAc+D+C_7H_{10}SO_3H \xrightleftharpoons{K_A} ROAc \cdots D+C_7H_{10}SO_3H$$

constant K_T derived from the kinetic data is in good agreement with the result of an independent spectroscopic determination. Detailed analysis of the thermodynamics of the solvolysis suggests the catalytic effect can be attributed to an increase in the entropy of activation (less negative $\Delta S^{\neq}$) by complex formation.

Catalysis of the racemization of optically active binaphthyl donors by organic acceptors has also been observed[77]. The donors were (+)-9,10-dihydro-3,4,5,6-dibenzophenanthrene (XVI) and (+)-1,1-binaphthyl (XVII).

(XVI)   (XVII)

Catalytic effects parallel to acceptor strength are observed for the acceptors 1,3,5-trinitrobenzene, picryl chloride and 2,4,7-trinitrofluorene.

(XVIII)

The intermediacy of a complex has been proposed to account for the high specificity of 2-hydroxy-5-nitrobenzyl bromide (XVIII) for tryptophan in a reaction with proteins[78]. There is no spectroscopic evidence for this complex at present.

10. Complexes of biological materials

In Table II are collected all the literature data on molecular complexes of small biological molecules which we have been able to locate. It would be superfluous to discuss in detail every case, since the table has been made fairly detailed. In particular, the wavelength at which the association constant was determined, or the maximum of new absorption is given, and compared with the absorption maximum of uncomplexed material. In a very large number of the cases listed in Table II a charge-transfer band is not observed and a donor–acceptor interaction is therefore questionable.

(a) Flavins

A great number of complexes of the flavin cofactors have been studied. The desire to implicate a donor–acceptor interaction in enzyme–cofactor binding is behind most of this work. This also holds true for the complexes of the NAD-type molecules to be considered in the next section. The topic of co-factor binding has been reviewed in detail by Shifrin and Kaplan[79].

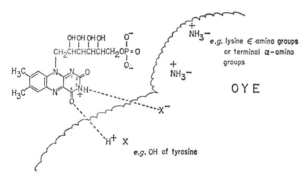

Fig. 18. Scheme for binding FMN to old yellow enzyme proposed by Theorell and Nygaard[80].

Insofar as the flavin complexes are concerned, the work in this field may be summarized as follows. About ten years ago, Theorell and Nygaard[80] proposed a scheme for the binding of FMN to the old yellow enzyme which envisioned hydrogen bonding at the isoalloxazine nucleus, as in Fig. 18. Tyrosine was implicated at the binding site by iodination studies.

Harbury and co-workers[81,82] attempted to test this suggestion by examin-

(Text continued on p. 128)

References p. 143

TABLE II MOLECULAR COMPLEXES OF BIOLOGICAL MATERIALS

Entry No.	Complex (a blank space in this column indicates that the component is the same as the preceding one)		Reference	Association constant[a]	Molar extinction coefficient[b]	Method[c]	λ (mμ)	Other
1		FMN	81	200–1000	Not reported	BH	445, 373	λ_{max} FMN 447, 375
		Riboflavin	81	200–1000	Not reported	BH	445, 373	
		3-Methyl-riboflavin	81	200–1000	Not reported	BH	445, 373	λ_{max} 3-Methyl-riboflavin 445, 373
2		FMN	81	100–200	Not reported	BH	445, 373	
		Riboflavin	81	100–200	Not reported	BH	445, 373	
		3-Methyl-riboflavin	81	100–200	Not reported	BH	445, 373	
3		FMN	81	100–200	Not reported	BH	445, 373	
4		FMN	81	100–200	Not reported	BH	445, 373	
5		FMN	81	20–100	Not reported	BH	445, 373	
6		FMN	81	20–100	Not reported	BH	445, 373	

Compound		Flavin	Ref.			Method	λ (nm)	Conditions
Tryptophan	—	FMN	81	20–100	Not reported	BH	445, 373	
8 Caffeine	—	FMN	87	60	Not reported	BH	500	See text, p. 129
	—	3-Methyl-riboflavin	81	20–100	Not reported	BH	445, 373	
	—	Riboflavin[d]	81					
	—	Riboflavin	109	Not determined	Not reported	BH	570, 620	1% HCl, ice, −70°
	—	FMN	81	20–100	Not reported	BH	445, 373	
	—	FMN[e]	104	90		Fluorescence		17°, pH 7.5
	—	FMN	104	103		Fluorescence		5°
	—	FMN	105	53		Fluorescence		
9 (CH₃O–C₆H₄–COO⁻)	—	Riboflavin	81	20–100	Not reported	BH	445 or 373	
	—	3-Methyl-riboflavin	81	20–100	Not reported	BH	445 or 373	
	—	FMN	81	20	Not reported	BH	445 or 373	
	—	Riboflavin	81	<20	Not reported	BH	445 or 373	
	—	3-Methyl-riboflavin	81	<20	Not reported	BH	445 or 373	
10 (HO–C₆H₄–COO⁻)	—	FMN	81	<20	Not reported	BH	445 or 373	
	—	Riboflavin	81	<20	Not reported	BH	445 or 373	
	—	3-Methyl-riboflavin	81	<20	Not reported	BH	445 or 373	

TABLE II (continued)

Entry No.	Complex (a blank space in this column indicates that the component is the same as the preceding one)		Reference	Association constant[a]	Molar extinction coefficient[b]	Method[c]	$\lambda\ (m\mu)$	Other
11	(salicylate, OH / COO⁻)	FMN	81	<20	Not reported	BH	445 or 373	
12	(salicylate, OH / COO⁻)	FAD	106	$1.54\cdot10^3$ (!)		Fluorescence		
13	(phenol, OH)	FMN	81	<20	Not reported	BH	445 or 375	
		Riboflavin	107	5		BH	445 or 375	
14	(chlorophenol, OH / Cl)	FMN	81	<20	Not reported	BH	445 or 375	
15	(pentachlorophenolate, Cl Cl Cl O⁻ Cl Cl)	Riboflavin[f]	82	455 (15°)	Not reported	BH	445 or 375	
16	(HO–C₆H₄–CH₂CHCOOC₂H₅, NH₂)	3-Methyl-riboflavin	81	<20	Not reported	BH	445 or 375	

No.	Compound		Flavin						Conditions
18	Tyrosine; Phenylalanine ($CH_2CHCOOH$, NH_2)	—	3-Methyl-riboflavin	81	<20	Not reported	BH	445 or 373	
		—	FMN	97	No significant spectral changes				
		—	Riboflavin	118	Not determined	Not determined	Difference spectrum		
19	(quinone)	—	FMN	81	<20	Not reported	BH	445 or 373	
20	(benzoate, COO^-)	—	FMN	81	<20	Not reported	BH	445 or 373	
		—	3-Methyl-riboflavin	81	<20	Not reported	BH	445 or 373	
		—	Riboflavin	81	<20	Not reported	BH	445 or 373	
21	Serotonin ($CH_2CH_2NH_2$, HO-)	—	FMN[g]	87	400	~4500	BH	500	
		—	FMN	109	Not determined	Not determined		570, 620	1% HCl, ice, −70°
22	Leucine	—	Riboflavin	118	Not determined	Not determined	Difference spectrum		No CT band
23	Glycine	—	Riboflavin	118	Not determined	Not determined	Difference spectrum		No CT band
24	$FMNH_2$	—	FMN	57	18	680 (900)	Indirect	900	
25	NADPH	—	FMN	109	Not determined	Not determined	Increased long λ absorption		In ice at −70°
26	NADH	—	FMN	109	Not determined	Not determined	Increased long λ absorption		In ice at −70°
		—	FMN	110	No evidence for complex		Fluorescence		Water, room temp.

TABLE II (continued)

Entry No.	Complex (a blank space in this column indicates that the component is the same as the preceding one)		Reference	Association constant[a]	Molar extinction coefficient[b]	Method[c]	λ (mμ)	Other
27	(1,2-naphthalenediol structure)	FMN	97[h]	242	Not reported	Modified Scott Eqn.	<500	pH 6.8
		FMN	97	68	Not reported	Modified Scott Eqn.	500	12 N HCl
28	(1,2-benzenediol structure)	FMN	97	10.4	Not reported	Modified Scott Eqn.	?	pH 6.8
		FMN	97	0.68	Not reported	Modified Scott Eqn.	?	12 N HCl
		Lumiflavin	97	2.9	Not reported	Modified Scott Eqn.	?	12 N HCl
		Riboflavin	97	3.2	Not reported	Modified Scott Eqn.	?	12 N HCl
		Riboflavin	97	3.9	Not reported	Modified Scott Eqn.	?	6 N HCl
29	(1,4-benzenediol structure)	Riboflavin	97	2.9	Not reported	Modified Scott Eqn.	?	6 N HCl
30	(trimethylbenzenediol structure)	Riboflavin	97	0.10	Not reported	Modified Scott Eqn.	?	6 N HCl
31	(1,3-benzenediol structure)	Riboflavin	97	3.3	Not reported	Modified Scott Eqn.	?	6 N HCl
32	(benzenediol structure)	Riboflavin	97	6.2	Not reported	Modified Scott Eqn.	?	6 N HCl

No.	Structure		Riboflavin						
34	(CH$_3$, OH; OH, H$_3$C, OH)	—	Riboflavin	97	6.2	Not reported	Modified Scott Eqn.	?	6 N HCl
35	(OH; HO, OH)	—	Riboflavin	97	7.4	Not reported	Modified Scott Eqn.	?	6 N HCl
36	(OH; CH$_3$, HO, OH)	—	Riboflavin	97	9.2	Not reported	Modified Scott Eqn.	?	6 N HCl
37	(OH, OH naphthalene)	—	Riboflavin	97	55	750	Modified Scott Eqn.	625	6 N HCl
38	(OH; HO, OH naphthalene)	—	Riboflavin	97	98	Not reported	Modified Scott Eqn.	550	6 N HCl
39	(OH, CH$_3$, CH$_3$; OH)	—	Riboflavin	97	55	Not reported	Modified Scott Eqn.	?	6 N HCl
40	(CH$_3$; OH, H$_3$C)	—	Riboflavin	97	88	Not reported	Modified Scott Eqn.	?	6 N HCl

TABLE II (continued)

Entry No.	Complex (a blank space in this column indicates that the component is the same as the preceding one)	Reference	Association constant[a]	Molar extinction coefficient[b]	Method[c]	λ (mμ)	Other
41	(naphthalenediol, OH / OH) — Riboflavin	97	111	Not reported	Modified Scott Eqn.	590	6 N HCl
42	(naphthalenediol, OH / HO) — Riboflavin	97	102	Not reported	Modified Scott Eqn.	<500	6 N HCl
43	Adenosine — Riboflavin	104	140		Fluorescence		5°
44	Adenosine — Riboflavin	104	120		Fluorescence		17°
	FAD (internal complex)	104	Not determined		Fluorescence		λ_{max} FMN 447, λ_{max} FAD ~450
	FAD (internal complex)	100	80% folded		Fluorescence		λ_{max} FAD ~450
45	(phenol, OH) — FAD	106	31	Not reported	BH	450	
46	(aminophenol, OH / NH$_2$) — FAD	106	9	Not reported	BH	450	
47	(H$_2$N– –COOH) — FAD	107	13	Not reported	BH	450	
48	(OH) — FAD	106	$2 \cdot 10^3$		Fluorescence		

No.	Structure		Substance			Method	
49		—	FAD	106	$5 \cdot 10^4$	Fluorescence	
50		—	FAD	106	$2.5 \cdot 10^4$	Fluorescence	
51		—	FAD	106	$8.3 \cdot 10^4$	Fluorescence	
52		—	FMN	98	Not determined	Not determined	Hypochromic effect	200–300
53		—	Riboflavin	100	17.2	Fluorescence	
54		—	Riboflavin	100	14.1	Fluorescence	
55		—	Riboflavin	100	13	Fluorescence	
56		—	Riboflavin	100	8.8	Fluorescence	

TABLE II (continued)

Entry No.	Complex (a blank space in this column indicates that the component is the same as the preceding one)	Reference	Association constant[a]	Molar extinction coefficient[b]	Method[c]	λ (mμ)	Other
57	Riboflavin —	100	7.6		Fluorescence		
58	Riboflavin —	100	7.6		Fluorescence		
59	Riboflavin —	100	7.9		Fluorescence		
60	Riboflavin —	100	6.5		Fluorescence		
61	Riboflavin —	100	6.4		Fluorescence		
62	Riboflavin —	100	3.9		Fluorescence		
	Riboflavin	100	2.8		Fluorescence		

65		—	Riboflavin	100	2.4	Fluorescence
66		—	Riboflavin	100	2.0	Fluorescence
67		—	Riboflavin	100	1.8	Fluorescence
68		—	Riboflavin	100	0.53	Fluorescence
69		—	Riboflavin	100	0.50	Fluorescence
70		—	Riboflavin	100	0.14	Fluorescence
71		—	Riboflavin	100	0.5	Fluorescence

TABLE II (continued)

Entry No.	Complex (a blank space in this column indicates that the component is the same as the preceding one)			Reference	Association constant[a]	Molar extinction coefficient[b]	Method[c]	λ (mμ)	Other
72		Riboflavin	—	100	0.18		Fluorescence		
73		Riboflavin	—	100	5.6		Fluorescence		
74		Riboflavin	—	98	Not determined	Not determined	Hypochromic effect	200–300	
75	FMNH₂		—	92	Not determined	Not determined	Difference spectrum	~510	
76	FMNH₂		—	92	Not determined	Not determined	Difference spectrum	~610	
77	FMNH₂	NAD⁺	—	92	idem	idem	idem	~580	
78	Lipoyl dehydrogenase–FADH₂	NAD⁺	—	91, 92	idem	idem	idem	720	
79	D-Amino acid oxidase–FAD	Δ¹,²-pyrroline 2-carboxylate[l]	—	91	idem	idem	idem	615	
80	D-Amino acid oxidase–FAD	Δ¹,²-piperidine 2-carboxylate[m]	—	91	idem	idem	idem	640	
81	D-Amino acid oxidase–FAD	indole 2-carboxylate	—	91	idem	idem	idem	530	
82	D-Amino acid oxidase–FAD·		—	91	idem	idem	idem	565	

No.	Compound		Ref.	Not determined	Not determined	Difference spectrum	spectrum	pH
84	D-Amino acid oxidase–FAD	—	91	Not determined	Not determined	Difference spectrum	<530	
85	D-Amino acid oxidase–FADH$_2$ — Δ^{12}-piperidine 2-carboxylate^{m,i}		91	idem	idem	idem		
86	Butyryl-CoA dehydrogenase (green)– FAD — unknown group		91	idem	idem	idem	720	
87	NADH–cytochrome b_5 reductase–FADH$_2$ — NAD		91	idem	idem	idem	Flat long wavelength absorption	
88	Glutathione reductase–FADH$_2$ —	—	91	idem	idem	idem	720	
89	NADP	—	102	2.5 2.2 2.47	540(370) 890(380) 560(370) 640(400)	BH		pH 6.6
90	Glycyl-L-tryptophan	—	102	2.9 2.98 2.92	500(400)	BH		pH 6.6–6.1
91		—	102	2.2 2.52 2.20	860(370) 540(390) 430(400)	BH		
92	Serotonin creatinine sulfate	—	102	1.8 1.81 2.07	1410(380)	BH		pH 6.4–6.6
93	Acetyltryptophan	—	102	4.0 5.0 4.73	510(400)	BH		pH 6.5
94		—	102	4.1 4.09 4.41	1220(370)	BH		pH 6.3–6.7
95	Yohimbine	—	102	1.42, 1.41	1220(370)	BH		pH 2.2–4.0

TABLE II (continued)

Entry No.	Complex (a blank space in this column indicates that the component is the same as the preceding one)	Reference	Association constant[a]	Molar extinction coefficient[b]	Method[c]	λ (mμ)	Other
96	tryptophan (CH₂CHCOOH, NH₂, indole) — CN-pyridinium, CH₂C₆H₅	102	4.9	890	BH		
97	COCH₃-pyridinium, CH₂C₆H₅	102	No reproducible results				
98	NAD (internal complex)	109	Not reported	Not reported			Ice, −70°
99	NADH (internal complex)	111	Not reported	Not reported	Fluorescence		
100	NADH (internal complex)	112	20–25	Not reported	Fluorescence		
101	Other internal complexes (?): Deamino - NAD, Deamino - NADH, 3-acetylpyridine - NAD, 3-pyridinealdehyde - NAD, 3-pyridinealdehyde - deamino - NAD	113			Fluorescence		
102	NADH — 1,3,5-trinitrobenzene (NO₂, NO₂, O₂N)	114	Not determined	Not determined	New absorption		
103	Phenothiazine — chloranil (Cl, Cl, Cl, Cl, O, O)	114	Not determined	Not determined	New absorption		In chloroform
	Phenothiazine	55	Not de-	Not de-	Infrared		Material isolated as solid from buffers

No.	Donor	Acceptor		Ref.			Observation	λ (nm)	Remarks
104	(name partly shown)	(nitro-substituted ring, structure cut)	—	114	Not determined	Not determined	New absorption		In chloroform
105	Phenothiazine	p-benzoquinone	—	114	Not determined	Not determined	New absorption		In chloroform
106	Indole	7,7,8,8-tetracyanoquinodimethane (TCNQ)	—	114	Not determined	Not determined	New absorption		Chloroform, observed two maxima (reaction?)
107	Indole-3-acetic acid (CH$_2$COOH-indole)	tetracyanoethylene, $(NC)_2C{=}C(CN)_2$	—	31	Not determined	Not determined	New absorption		CH$_2$Cl$_2$, complex formation followed by reaction
108	Adenine, caffeine	1,3-dinitrobenzene	—	114	Not determined	Not determined			
109	Adenosine	chloranil (tetrachloro-p-benzoquinone)	—	114	Not determined	Not determined			
110		—	—	115	Not determined	Not determined	Color change	480[1]	Intensity reaches maximum after 15 min
111	Guanosine	—	—	115	Not determined	Not determined	Color change	610[1]	Intensity reaches maximum after 8 h
112	Deoxycytidine · HCl	—	—	115	Not determined	Not determined	Color change	Badly[1] defined maximum	

TABLE II (continued)

Entry No.	Complex (a blank space in this column indicates that the component is the same as the preceding one)	Reference	Association constant[a]	Molar extinction coefficient[b]	Method[c]	λ (mμ)	Other
113	Thymidine	—	115	Not determined	Not determined	Color change	Badly defined maximum	
114	Adenylic acid	—	115	Not determined	Not determined	Color change	500[i]	Intensity reaches maximum after 2 h
115	Guanylic acid	—	115	Not determined	Not determined	Color change	570[i]	Intensity reaches maximum after 6 h
116	Deoxycytosine 5'-phosphate	—	115	Not determined	Not determined	Color change	Badly[i] defined maximum	
117	Thymidine 5'-phosphate	—	115	Not determined	Not determined	Color change	550[i]	
118	CH$_2$CH–COOH NH$_2$ (indole structure) Acridine	—	117	23	$6.7 \cdot 10^2$ (390)[p]	BH	390	
119	HO–(indole structure) CH$_2$CHCOOH NH$_2$ Acridine	—	117	100	$-1.1 \cdot 10^3$ (390)[p]	BH	390	
120	(acridine structure)	—	116	Not determined	Weak	New absorption	~460[j]	
121	(structure with NH$_2$, H$_2$N)	—	116	Not determined	Strong	New absorption	~528[j]	
122	(structure with NH$_2$)							

No.	Compound		Reagent		termined	termined	New peak / BH	change during 3 days
124	Thymidilic acid	—	H_2N–...–N–...–NH_2	121	*idem*	*idem*	New peak	560[k] *idem*
125	Deoxyribonucleic acid	—		121	*idem*	*idem*	New peak	506[k] Opacity of solution increases during 2 days, but no detectable color change 1000
126	β-Carotene	—	I_2	56	*idem*	*idem*		
127	Numerous carcinogenic hydrocarbons	—	I_2	120	*idem*	*idem*		
128	Tryptophan, proline, glycine, alanine	—	O_2	108	*idem*	*idem*	New absorption	350–410[r]
129	HO–⟨C₆H₄⟩–$CH_2CHCOOH$ (NH_2)	—	chloranil ($Cl_4C_6O_2$)	119[n]	*idem*	*idem*		360 λ_{max} chloranil 285, 380
130	$(CH_3)_2CH$–$CH_2CHCOOH$ (NH_2)	—		119	224	Not determined	BH	360 pH 7
131	$CH_3CHCOOH$ (NH_2)	—		119	318	Not determined	BH	350 pH 8
132	indolyl–CH_2–$CHCOOH$ (NH_2)	—		119	176	Not determined	BH	355 $k\sim0$ at pH 12
133	H_2NCH_2COOH	—		119	298	Not determined	BH	390
134	⟨C₆H₅⟩–$CH_2CHCOOH$ (NH_2)	—		119	Not determined	Not determined		360 pH 7.4

TABLE II (continued)

Entry No.	Complex (a blank space in this column indicates that the component is the same as the preceding one)	Reference	Association constant[a]	Molar extinction coefficient[b]	Method[c]	λ (mμ)	Other
135	[structure: ring with –COOH and N–H] —	119	Not determined	Not determined		360	pH 8
136	Amethopterin — Tryptophan	117	45	$1.6 \cdot 10^2$ p	BH	420	Pteridines as a class have a maximum in the vicinity of 380mμ
137	Amethopterin — Serotonin	117	60	$1.6 \cdot 10^2$ p	BH	420	
138	Aminopterin — Tryptophan	117	25	$5.2 \cdot 10^2$ p	BH	430	
139	Aminopterin — Serotonin	117	45	$5.2 \cdot 10^2$ p	BH	430	
140	N10-Methylfolic acid — Tryptophan	117	22	$3.6 \cdot 10^2$ p	BH	420	
141	N10-Methylfolic acid — Serotonin	117	34	$3.6 \cdot 10^2$ p	BH	420	
142	Folic acid — Tryptophan	117	13	$1.1 \cdot 10^3$ p	BH	400	
143	Folic acid — Serotonin	117	30	$1.1 \cdot 10^3$ p	BH	400	
144	N10-Formylfolic acid — Tryptophan	117	15	$7.1 \cdot 10^2$ p	BH	400	
145	N10-Formylfolic acid — Serotonin	117	23	$7.1 \cdot 10^2$ p	BH	400	
146	9-Methylfolic acid — Tryptophan	117	9	$8.5 \cdot 10^2$ p	BH	400	
147	9-Methylfolic acid — Serotonin	117	20	$8.5 \cdot 10^2$ p	BH	400	
148	2,4-Diamino-6,7-dimethylpteridine — Tryptophan	117	7	$-1.2 \cdot 10^3$ p	BH	370	
149	2,4-Diamino-6,7-dimethylpteridine — Serotonin	117	18	$-1.2 \cdot 10^3$ p	BH	370	
150	2-Amino-4-hydroxypteridine 6-carboxylic acid — Tryptophan	117	5	$2.2 \cdot 10^3$ p	BH	410	
151	2-Amino-4-hydroxypteridine 6-carboxylic acid — Serotonin	117	15	$2.2 \cdot 10^3$ p	BH	410	
152	Xanthopterin — Tryptophan	117	2	$5.0 \cdot 10^3$ p	BH	410	

No.	Compound	Interactant		Page			Remarks
155	(indole) $CH_2CH\text{-}COOH$, NH_2	I_2	—	122	Not determined	Not determined	shows sharp EPR signal
							idem
156	3-Methylcholanthrene	Vitamin K_3	—	123	Not determined	Not determined	Red complex, in CH_2Cl_2, λ_{max} Vit. K_3 325mμ
157	1,2-Dibenzanthracene	Vitamin K_3	—	123	Not determined	Not determined	480–490j
158	3,4-Benzpyrene	FMN	—	124	Not determined	Not determined	480–490j
							490j
159	Several carcinogenic hydrocarbons	Substituted benzoquinones	—	126	Not determined	Not determined	Fusionq point of mixture
160	Chlorpromazine	FMN	—	138	Not determined	Not determined	Color change Ice, −70°

a For uniformity, all equilibrium data are presented as an association constant. If originally reported as a dissociation constant, the reciprocal is given here. Units are liter/mole.

b Units are liter/mole·cm. Numbers in brackets following the constant indicate the wavelength.

c BH refers to the Benesi–Hildebrand Eqn. (2). We use this designation if the equation in the original paper is of this form, although the authors may not have designated it as such.

d ΔH of complex formation reported as -6.1 $kcal \cdot mole^{-1}$ at pH 6.9. Shifts in riboflavin redox potential in presence of interactant also noted.

e ΔH of complex formation reported as -1.6 ± 0.6 $kcal \cdot mole^{-1}$.

f ΔH of complex formation reported as -3.3 $kcal \cdot mole^{-1}$ at pH 9.2.

g The comments in the text concerning the tryptophan–FMN complex presumably apply also to this complex.

h All experiments reported in ref. 97 where 6 N HCl is indicated, the solvent was also 50% in ethanol.

i Maximum of new absorption, considerable absorption on long wavelength side of maximum, solvent is dimethylsulfoxide.

j Maximum of new absorption.

k Maximum of new peak appearing as a shoulder.

l Product of enzymatic oxidation of D-proline.

m Product of enzymatic oxidation of D-pipecolic acid.

n All experiments reported in ref. 119 are in 50% aqueous ethanol.

p Extinction reported as $\varepsilon_{mixture} - \varepsilon_{components}$ at the indicated wavelength.

q No correlation between quinone reduction potential or hydrocarbon ionization potential and complex stability.

r Fine structure observed in the CT band with tryptophan attributed to tryptophan singlet to triplet transition.

ing in detail a number of phenol complexes of riboflavin and 3-methyl-riboflavin, where hydrogen bonding at position 3, as in Fig. 18, would be impossible. Finding that complexes formed equally well with both methylated and unmethylated flavin they were inclined to believe that charge-transfer forces, not hydrogen bonding, might be more predominant in these interactions.

It perhaps adds perspective to recall what the situation regarding such interactions was at that time. The "hydrophobic" bond concept[83], which gave to water an important role in such interactions, was not well established. Mulliken, in an early classic paper on charge transfer, had suggested (in one sentence) that charge transfer "might afford new possibilities for understanding intermolecular interactions in biological systems". Kosower[84,85] had already observed the charge-transfer absorption of several pyridinium salts in the nicotinamide series and had suggested that a donor–acceptor complex could account for the 360-mμ absorption in triosephosphate dehydrogenase (EC 1.2.1.12). In this context Harbury's suggestion and (see below) that of Isenberg and Szent-Gyorgyi were completely reasonable.

There is no doubt that many of the complexes studied by Harbury and co-workers exist, but there is serious question as to whether they are of the donor–acceptor type. The spectrum of one of the complexes, typical of them all, is reproduced in Fig. 19. The complex with naphthoate ion indicated there

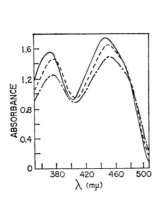

Fig. 19. Absorption spectra of 3-methyl-riboflavin plus - - - -, 0.21 M benzoate; —·—·—·, 0.02 M naphthoate.

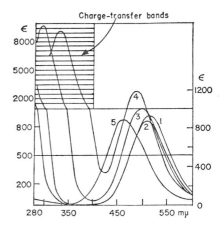

Fig. 20. Charge-transfer band of donor-acceptor complexes of iodine with various solvent molecules. 1, CCl_4; 2, $C_6H_5CF_3$; 3, C_6H_6; 4, $C_6H_3(CH_3)_3$; 5, $(C_2H_5)_2O$.

was one of the strongest complexes in their series. The effect of phenol is less pronounced, as is that of tyrosine. It seems certain that the hypochromicity and small red shift are not due to charge-transfer absorption. Compare the spectrum in Fig. 19 with the situation encountered by Benesi and Hildebrand[86] and which prompted Mulliken to undertake a theoretical study (Fig. 20). The charge-transfer band in Fig. 20 is definite and unmistakable.

At about the same time as Harbury's work appeared, Isenberg and Szent-Gyorgyi[87] reported a study of the red complexes of FMN with tryptophan and serotonin (Table II, entries 7, 21). The difference spectrum of complexed *versus* uncomplexed FMN showed a maximum at 503 mμ, comparable to the maximum at 500 mμ known for riboflavin semiquinone in acid solution. On this spectral evidence, it was proposed that an electron transfer had occurred to give riboflavin semiquinone. That semiquinone formation had occurred was questioned almost immediately[82], but theoretical calculations[89] indicated that riboflavin could indeed be a good acceptor. Later[96] an ESR signal characteristic of flavin semiquinone was observed in an acidic solution of the complexes of FMN with serotonin and tryptophan. This indicated that electron transfer could occur under appropriate conditions, but no signal was observed in neutral solutions. The weak signal observed in acid solution with tryptophan as donor now appears to have been due to the influence of light, however[88].

TABLE III

CHARGE-TRANSFER BANDS OF INDOLE COMPLEXES[a]

	Max. (mμ)	E_{CT} (kcal mole^{-1})
(1) FMNH$_2$: FMN	900	31.8
(2) FMNH$_2$: NAD$^+$	700	40.8
(3) Indole : chloranil	ca. 500	53.0
(4) Indole : FMN	330 (estimate)	86.3 (estimate)
(5) Indole : NAD$^+$	ca. 300	95.3

[a] Data from Kosower, ref. 90.

Here the matter has rested, but recently Kosower[90] has raised a cogent point. He notes that the charge-transfer transition from indole to riboflavin should appear at about 330 mμ, not at 500 mμ in the visible region of the spectrum. The basis of this suggestion lies in an estimate of the relative electron affinity of riboflavin in the series in Table III. The details of Kosower's analysis are not available to us at present. One can, however, use the molecular orbital energies calculated by the Pullmans to make a rough estimate of $I_p - E_a$ which, as we have already noted, differs from the charge-transfer energy only by a perturbation term which may be reasonably constant. These estimates, made using the orbital energies given in the Pullmans' book[11], are given in Table IV with the experimentally known charge-transfer energies. The calculated $I_p - E_a$ in units of β, the resonance integral of MO theory, (a negative energy quantity) reproduce the trends in experimental energy values quite well. A plot of ΔE_{CT} versus $I_p - I_a$ shows some scattering of points but does support an estimate of 80–90 kcal for the charge-transfer energy from indole to FMN. This is also quite evident from examination of the $I_p - I_a$ entries in Table IV for the indole–NAD$^+$ and indole–FMN complexes. While the estimate is rough, it seems unlikely to be in

TABLE IV

ESTIMATE OF CHARGE-TRANSFER ENERGY FROM MOLECULAR-ORBITAL
PARAMETERS

	I_p	E_a	$I_p - E_a{}^a$	E_{CT}
FMNH$_2$	$+0.105\,\beta$		$\sim -0.24\,\beta$	31.8 kcal
FMN		$-0.344\,\beta$		
FMNH$_2$	$+0.105\,\beta$		$\sim -0.25\,\beta$	40.8 kcal
NAD$^+$		$-0.356\,\beta$		
Indole	$-0.534\,\beta$		$-0.63\pm0.05\,\beta$	53.0 kcal^c
Chloranil		$-0.1\pm0.05\,\beta^b$		
Indole	$-0.534\,\beta$		$\sim -0.88\,\beta$	80–90 kcal
FMN		$-0.344\,\beta$		(estimate)
Indole	$-0.534\,\beta$		$\sim -0.89\,\beta$	86–95 kcal^d
NAD$^+$		$-0.356\,\beta$		

^{a} The algebraic signs here are tricky. This is ΔH. Since by definition $\Delta H = I_p$, but $\Delta H = -E_a$, algebraic addition of the terms in the table gives the correct result. Since β is negative, the total ΔH is positive, as expected.

^{b} The Pullmans quote $-0.23\,\beta$ for benzoquinone, $-0.18\,\beta$ for cyanoquinone. As chloranil is a much better acceptor than either, an estimate of -0.1 ± 0.05 is reasonable and conservative.

^{c} λ_{max} 355 mμ quoted in Table II (entry 132) for tryptophan–chloranil is for donation from the amino group, not the indole nucleus.

^{d} Using Kosower's estimate of $ca.$ 300 mμ or calculating from Fig. 26, using $\lambda_{max\ CT}$ 320 mμ.

error by the 25 or so kcal needed to save the former 500-mμ assignment for the charge-transfer band of the tryptophan–FMN complex. As no ESR signal is observed in a neutral solution of riboflavin and tryptophan[88,96], it now appears that the long wavelength absorption is due to broadening or splitting of the riboflavin absorption band located at 447 mμ.

Massey and co-workers have recently assigned certain absorption maxima in a number of flavoproteins (entries 78–88, Table II) to charge-transfer transitions[91]. Conditions favoring reduced enzyme favor the appearance of the 720-mμ absorption in lipoyl dehydrogenase (entry 78). The appearance of this peak is attributed to formation of a complex involving FADH$_2$ and NAD$^+$. In this connection complexes of reduced riboflavin, N-methyl-3-nicotinamide and N-methyl-4-nicotinamide have been studied (entries 75, 76, Table II)[92]. The difference spectrum of one of these complexes is given in Fig. 21. The peak at 510 mμ is assigned to the charge-transfer band.

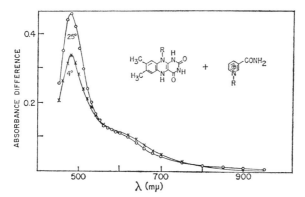

Fig. 21. Difference spectrum of FMNH$_2$ and NAD$^+$ relative to FMNH$_2$. (From ref. 95)

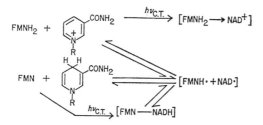

Fig. 22. Possible equilibrium relationships between oxidized and reduced forms of flavin and pyridine nucleotide coenzymes.

The green color of these complexes of reduced riboflavin, attributable to the long absorption tail, is reminiscent of the complex reported by Mahler and Brand[93]. This latter complex is prepared by grinding together solid riboflavin and NADH. An ESR signal is observed in the complex and its green color is likely attributable, in this instance, to riboflavin semiquinone. A relationship between the two series of FMN–NAD complexes such as that shown in Fig. 22 seems possible. A charge-transfer transition NADH→FMN is energetically reasonable, but chemical reaction between the materials would make it difficult to study and it has not been observed thus far.

Concerning lipoyl dehydrogenase, Kosower[4] has recently suggested that the red intermediate observed during the reduction of this enzyme is a complex of mercaptide and FAD (XIX). Searls and Sanadi[94] claimed to have detected a similar complex in a reaction between dihydrothioctate (XX) and FMN. An intensification of the orange tinge of the solution was observed immediately

after mixing the reagents. The difference spectrum of this solution *versus* FMN showed a maximum at 535 mμ which was assigned to a charge-transfer complex between thiol and FMN. With time the solution turned olive green and an ESR signal appeared. Massey and Atherton[95] were unable to reproduce this result under anaerobic conditions, although a reduction of FMN apparently did occur. They concluded that the 535-mμ absorption appeared only in the presence of light and some hydrogen peroxide which had been generated by air oxidation of reduced flavin.

Lumiflavin   Lumichrome

In the following we consider some of the more recent reports of complexes involving riboflavin. Fleishman and Tollin[88,97] have attempted to determine whether the acceptor strength of riboflavin can be increased by protonation of the isoalloxazine nucleus. They report complex formation between a number of phenols and riboflavin, and a few with lumiflavin and lumichrome, in 6–12 N acid (entries 27–42, Table II). In some cases highly colored complexes were isolated from solution. Discrete new absorption maxima, assigned to charge-transfer transitions, appear with naphthalenediols and trimethylquinol as donor (Fig. 23). With other phenols color changes are observed, since the riboflavin absorption broadens considerably out towards the green, but no discrete new maxima appear. Phenols substituted with electron-withdrawing groups (*i.e.*, tetrachloroquinol) show no color reaction. Interestingly, no color change is observed with 2,6-di-*tert*.-butylphenol. In answer to their original question, they find the stabilities of the complexes are actually lower in strong acid, but they are unable to correlate stabilities with any donor parameters. Surprisingly, an ESR signal, characteristic of ribo-

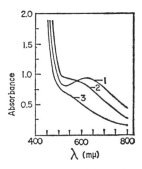

Fig. 23. 9-Methylisoalloxazine with phenols. Solvent is 50% 12 N HCl–50% ethanol. Flavin is 0.1 M, phenol 0.2 M. 1, 1,4-naphthalenediol; 2, 1,2-naphthalenediol; 3, trimethylquinol. (From ref. 88)

flavin semiquinone and increasing in intensity with increasing acid concentration, is observed[97]. No signal attributable to phenol radical is detected, presumably because of its disproportionation. From the results it seems that although complex stability decreases in acid solutions, electron-transfer reactions occur in acid solution, presumably due to the added stability of the flavin radicals under these conditions. Fleishman and Tollin also report[27] that their riboflavin hydroiodide complex forms stable highly colored complexes with phenols.

It is also interesting to recall, in connection with flavin complexes in acid solution, another result of Isenberg and Szent-Gyorgyi[109]. In 1% aqueous HCl, solutions of serotonin and indole with FMN show a new absorption at 570 mμ and a shoulder at 620 mμ when the spectra are observed at low temperatures. This absorption is at considerably longer wavelength than they observed in neutral solutions.

A complex of 1,1,3-tricyano-2-aminopropene (XXI) and riboflavin (entry 74, Table II) has been suggested to be of the charge-transfer type[98]. Such a complex would be of interest in that (XXI) has been reported to block oxidative phosphorylation[99]. A hypochromic effect is observed for the riboflavin maxima, but no new peak appears. The same note refers to charge-transfer complexes of tryptophan (a weak donor) and picric acid (a strong acceptor) with riboflavin. As one would anticipate a healthy contribution from dipolar structure (XXII) in the tricyanoaminopropene, it is not clear whether this compound should be a donor or acceptor (or alkylating agent). A donor–acceptor interpretation for the complex of (XXI) and riboflavin seems questionable.

A proposal has been put forward[100] that "electrostatic" forces can play a role in complexes of a number of purines and pyrimidines (entries 50–69, Table II) with riboflavin. The interaction is suggested to be favored by the complementarity of charge (obtained as charge densities in Hückel calculations) in the regions C-6 to N-9 in the purines and N-1 to N-10 of the flavin (Fig. 24). These fractional charges are manifestations of the slight polarity of

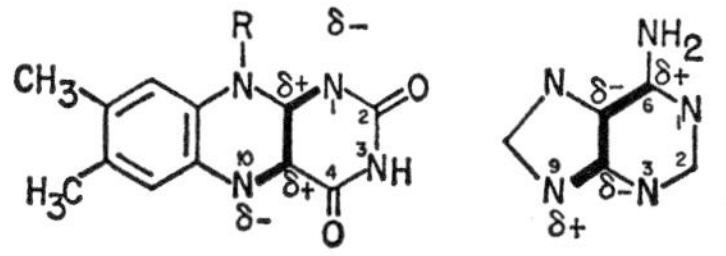

Fig. 24. Supposed pairing of charges in overlap of adenine and isoalloxazine.

the individual bonds and arise from slight polarization of the bond due to the different electronegativities of the component atoms. Whether the interaction is equivalent to interaction between oriented dipoles will depend on the intermolecular distance in the complex. The question of the applicability of a dipole–dipole or electrostatic model has been discussed, for example, by Hirshfelder[100a]. Partial transfer of an electron from donor to acceptor is viewed as enhancing the interaction by increasing the charge on oppositely paired atoms. This proposal is along the lines previously suggested by Karreman[101] for the tryptophan–riboflavin complex. No charge-transfer band is observed in any of the above complexes, only the usual hypochromic effect in the riboflavin absorption. In dimethylformamide, the complexes are barely demonstrable even by fluorescence techniques, which leads one to suspect a contribution from water in stabilizing the complexes. Straightforward calculation of charge densities in this way probably overestimates the effects, if there are any, in that contributions from σ electrons are neglected. Polarization of σ electrons would be opposite to the π electron polarization.

Finally, we call attention to the remarkable stability of the complexes of nitrophenols[106] and FAD (entries 48–51, Table II). This stability may be due to protonation of the adenine moiety of FAD by the phenols. Briegleb and Delle[106a] have detected salt formation in complexes of aromatic amines and picric acid (pK_a $1.8 \cdot 10^{-1}$). They used infrared absorption spectroscopy and detected a typical salt band at 3.4 μ (2910 cm^{-1}).

(b) Pyridine nucleotides

The overall course of development of work in this area parallels that of the flavin series.

Following the lead of Kosower who had established the identity of the charge-transfer bands in the pyridinium iodides, Cilento and Guisti[102] sought complexes of a number of indoles with NAD$^+$ and its analogues as possible models for enzyme binding. Similar studies have been reported by Alivasitos et al.[125]. The broad new absorption appearing in the spectrum of the tryptophan–NAD$^+$ mixture is shown in Fig. 25. It should be noted that the small association constants reported in this series (entries 89–96, Table II) are in the range of uncertain reliability of the Benesi–Hildebrand method.

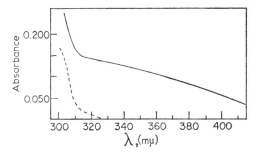

Fig. 25. Absorption spectrum of L-tryptophan in the presence of NAD$^+$. $1.71 \cdot 10^{-2}$ M tryptophan + 56.5 mg NAD$^+$/ml. Dashed line is tryptophan absorption. (From ref. 102)

Shifrin[103] has tried a novel approach to the problem. He has synthesized two series of molecules in which the potential donor and acceptor sites are in the same molecule (see XXIII and XXIV). Both absorption and emission

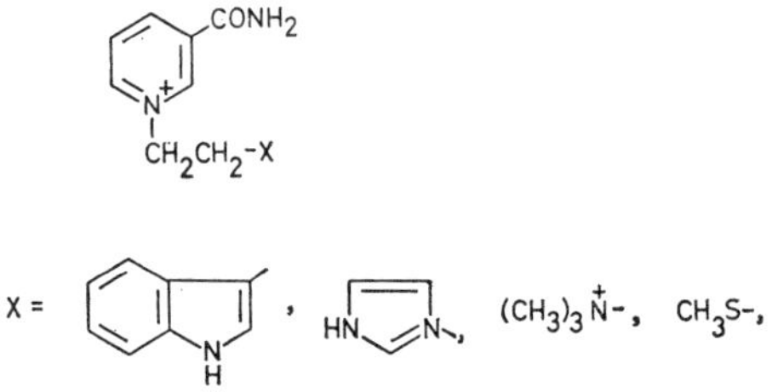

(XXIII)

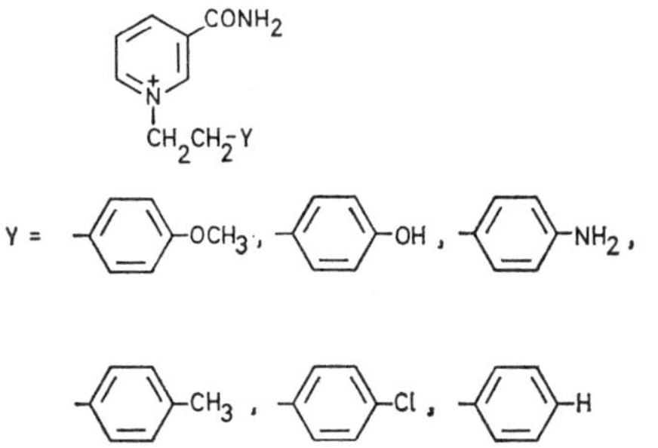

(XXIV)

spectra were studied for the series (XXIII) in an attempt to reproduce the spectral behavior of bound NAD^+. The enhanced fluorescence of enzyme-bound NAD^+ was not reproduced in these models. The absorption spectra of all the above compounds show absorption tails on the long wavelength side of the maxima (Fig. 26). These absorption tails are assigned to a charge-transfer transition, but a case such as that in trace C of the figure may be

Fig. 26. Trace A, absorption spectrum of a methanolic solution of indolylethylnicotinamide. Comparison with Fig. 25 reveals similarities in the 320–420 mμ region. Trace B, spectrum of N-(β-p-hydroxyphenylethyl)-3-carbamoylpyridinium chloride in water (solid curve) and that of an equimolar mixture of tyramine hydrochloride and N-methylnicotinamide perchlorate in water (dashed curve). Trace C, spectrum of N-(β-4'-imidazolylethyl)-3-carbamoylpyridinium chloride in methanol (solid curve). Dashed curve is the spectrum obtained subtracting absorption of N-methylnicotinamide perchlorate. (From ref. 103)

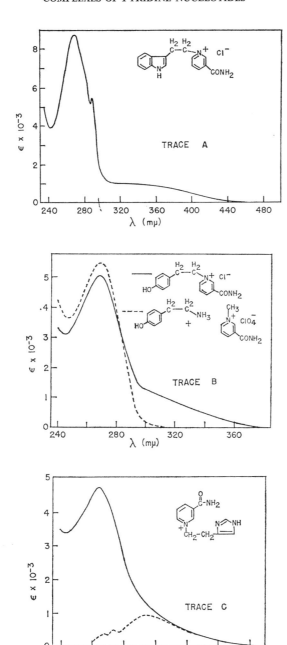

questionable. For the series (XXIV) the maxima, taken from difference spectra, can be correlated with the Hammett sigma values for the substituent[103a].

(c) Other complexes

We now turn to a consideration of complexes of other biological materials which have on occasion been linked with the topic of charge transfer. We defer a critical discussion of the possible physiological significance of the various complexes to specialists, and consider only the physical aspects of complex formation.

Complexes of various carcinogenic hydrocarbons with purines have long been of interest in the study of chemical carcinogenesis. From time to time theories have been proposed which invoke charge transfer or electron transfer. These various theories have recently been discussed by Mme. Pullman[127]. That purines can increase the solubility of polycyclic aromatic hydrocarbons has been appreciated for many years[128] and quantitative studies have been reported[129,130]. An X-ray crystallographic study[131] of a 1 : 1 crystalline complex between tetramethyluric acid and pyrene shows a face-to-face alternate stacking of the purine and hydrocarbon molecules. The interactions in solution similarly appear to involve a stacking-type association between partners, but for DNA–hydrocarbon association the intercalation versus adsorption question is still not settled[131a]. Some time ago[89] it was noted that the relative solubilizing power of the purines could be correlated with the energy of the highest occupied molecular orbital. This type correlation has often suggested a charge-transfer contribution to the intermolecular binding forces. However, no charge-transfer band is observed in these complexes and there has always remained a question as to the relative contribution of the charge-transfer energy to the overall binding. We note with interest a recent theoretical study of the intermolecular forces between purines and 3,4-benzpyrene[132]. By considering only classical dipole–induced dipole and London forces it was shown that the relative solubilizing power of the purines toward this hydrocarbon could be accounted for. An approximate expression for estimating London forces is obtained from a second-order perturbation treatment[133] and is:

$$W'' = -\frac{3}{2} \cdot \frac{\alpha_1 \alpha_2}{R^6} \cdot \frac{I_1 I_2}{I_1 + I_2}$$

where α_1, α_2 are the respective molecular polarizabilities, I_1, I_2 are the ionization potentials of the partners, and R is the intermolecular distance. The calculations are preliminary and have incorporated a number of approximations but show that a very large contribution from London forces is to be anticipated. The relative contribution from charge-transfer forces remains to be assessed, but one wonders if any contribution is to be found, particularly in the absence of a charge-transfer band.

A problem somewhat related to the above involves the binding of the mutagenic acridine dyes to nucleic acids. Considerable experimental effort has been devoted to distinguishing between an ionic association at the phosphate groups of the polymeric chain[134] and an intercalation of dye between base pairs of the chain[135]. Lerman[136] has recently reported a study of the chemical reactivity (diazotization) of the amino group in bound and unbound 5-aminoacridine. He finds the rate of reaction significantly decreased when the dye is bound to DNA and interprets this as due to an increased shielding of the amino group on intercalation of the dye between base pairs. An attempt to settle the matter by absorption spectroscopy has also been reported[121]. The reasoning was that if the dye were intercalated between bases there would be a considerable interaction with the π systems of the bases since the interannular distance would be about 3.5 Å. Under such circumstances one might find a charge-transfer absorption. For acriflavin bound to DNA, and proflavin to guanylic acid and thymidylic acid, a new shoulder appears in the dye-absorption band which is assigned to a charge-transfer transition with dye as acceptor (entries 123–125, Table II). This is offered as support for Lerman's intercalation mechanism. We note, however, that for the moderate donor thymine (highest occupied m.o. 0.510 β) and moderate acceptor proflavin (lowest empty m.o. -0.408 β) a charge-transfer energy as low as 50–55 kcal is quite surprising as is the fact that a similar complex of guanine (highest occupied m.o. 0.307 β) shows virtually the same maximum.

Calculations have recently revealed a correlation between the highest occupied molecular orbital and the hallucinogenic activity of a series of drugs of the LSD type[137]. This has caused some further speculation as to a possible role for charge transfer in the mode of action of these drugs. The anticipated donor properties of LSD (a substituted indole) and chlorpromazine[138] (XXV) have been noted previously and a possible role for charge transfer in their mode of action has been considered. Actinomycin C_3 (XXVI), a highly toxic antibiotic, is expected to have good acceptor properties[139].

Polycyclic aromatic hydrocarbons inhibit various flavin-containing enzymes[140,124]. This has been attributed to formation of donor–acceptor complexes. It is of interest that the absorption maximum reported for the complex of 3,4-benzpyrene (energy highest occupied m.o. 0.371 β) and FMN[124] (entry 158, Table II) corresponds roughly with that expected from the results of Table IV. The existence of complexes of polycyclic aromatic

hydrocarbons with FMN and quinones (*cf.* entry 159, Table II) has led to speculation that such complexes may play a role in certain aspects of chemical carcinogenesis[141], but this is a controversial point.

11. Complexes of metal cations

Charge-transfer transitions in complexed metal ions have been well documented[142,143]. An example is the case of the hydrated ferric ion, where a new absorption corresponding to a "metal-reduction spectrum" occurs at 330 mμ. Another is the 227.5-mμ absorption in the complex $[Co(NH_3)_5Cl]^{2+}$ attributed to electron transfer from chloride to metal ion. At low tempera-

$$Fe^{3+}(OH_2) \xrightarrow[330m\mu]{h\nu_{CT}} Fe^{2+} \cdots \overset{+}{O}H_2$$

tures, charge transfer to a frozen matrix can be made to reverse quite slowly. For example, ultraviolet irradiation of a frozen aqueous matrix of Fe^{3+} or I^- at 77° K gives an ESR signal characteristic of the hydrogen atom (a doublet with 500-gauss splitting), arising from electron transfer to or from the matrix[144].

An intense new absorption band occurs in a number of nickel–quinone complexes[145] such as (XXVII). This has been characterized as a charge transfer band. It might be noted that an iron complex of coenzyme Q (XXVIII) has been proposed[146] as an intermediate in oxidative phosphoryl-

O $=$ ... $=$ O
Ni
O $=$ ... $=$ O

(XXVII)

OPO$_3$H

O ... O
Fe^{2+}
O ... O

Enzyme
(XXVIII)

ation. As there is no biochemical evidence bearing on this proposal as yet, it seems premature to comment on the possible spectroscopic properties of such a complex.

Good evidence for charge-transfer transitions involving metal ions complexed to biological molecules seems to be rare. Orgel[147] has pointed out that the larger than ordinary extinction for the 575-mμ band of oxygenated hemocyanin may be due to a band of Cu^{2+} modified by a charge-transfer component. The copper-containing enzyme laccase has an unusually intense absorption in the vicinity of 600 mμ, possibly due to charge transfer to oxygen or enzyme[148]. Similarly, the blue color of ascorbic acid oxidase in the presence of oxygen[149] may be due to a charge-transfer transition. Charge-transfer contributions may also be involved in the long-wavelength absorption of some metal-containing porphyrins[150].

The increased absorption above 500 mμ observed for riboflavin in the presence of a number of transition metals[151] has been confirmed and extended by Radda and Calvin[110] who also noted a similar effect with sodium and magnesium ions. This is likely the result of a perturbation of molecular-orbital energies by the electrostatic effects of a coordinated metal ion, rather than a charge-transfer phenomenon. The shift of the long-wavelength absorption maximum of riboflavin from 447 mμ to 430 mμ in the presence of ferrous ion in a pyrophosphate, but not a phosphate or maleate buffer[152], is unexplained.

The bright red complex of silver ion and riboflavin[104] has been considered to involve charge transfer from metal ion to flavin (XXIX)[153]. This silver–riboflavin complex was the first of a remarkable series of "charge-transfer

(XXIX)

chelates" of riboflavin semiquinone which have been studied in some detail by Hemmerich and his co-workers[154]. With valence stable d metal ions at physiological pH, the equilibrium between oxidized and reduced flavin can

(XXX)

be shifted toward chelated radical semiquinone (termed a "comproportionation"). With coordinated metals such as Mo^V, Fe^{2+} and Cu^I (and Ag^I already referred to above) in acetonitrile solution, one can observe a redox

$$Fl_{ox}H + \underset{\text{reduced}}{FlH_2} + 2\,Me^{2+} \rightleftharpoons 2\,[FlHMe]^{+\cdot} + H^+$$

reaction and formation of the chelates (XXX). Redox processes of this type may play a role in redox catalysis in metal-containing flavoproteins[155].

REFERENCES

1 G. BRIEGLEB, *Electronen–Donator–Acceptor Komplexe*, Springer, Berlin, 1961.
2 R. S. MULLIKEN AND W. E. PERSON, *Ann. Rev. Phys. Chem.*, 14 (1963) 107.
3 L. J. ANDREWS AND R. M. KEEFER, *Molecular Complexes in Organic Chemistry*, Holden-Day, San Francisco, 1964.
4 E. M. KOSOWER, *Progress in Physical-Organic Chemistry*, Vol. III, Wiley, New York, 1965, p. 81.
5 A. SZENT-GYORGYI, *Introduction to a Submolecular Biology*, Academic Press, New York, 1960.
6 R. S. MULLIKEN AND W. E. PERSON, Ref. 2, footnote 5, p. 110.
7 R. S. MULLIKEN, *J. Am. Chem. Soc.*, 72 (1950) 600; *J. Chem. Phys.*, 19 (1951) 514; *J. Am. Chem. Soc.*, 74 (1952) 811; *J. Phys. Chem.*, 56 (1952) 801.
8 *Cf.* Ref. 1, p. 10.
9 M. J. S. DEWAR AND A. R. LEPLEY, *J. Am. Chem. Soc.*, 83 (1961) 4560.
10 H. DeVOE, *J. Chem. Phys.*, 41 (1964) 393, and references therein.
11 B. PULLMAN AND A. PULLMAN, *Quantum Biochemistry*, Wiley, New York, 1963.
12 R. E. MERRIFIELD AND W. D. PHILLIPS, *J. Am. Chem. Soc.*, 80 (1958) 2778.
13 R. BHATTACHARYA AND S. BASU, *Trans. Faraday Soc.*, 54 (1958) 1286.
14 M. E. PEOVER, *Trans. Faraday Soc.*, 58 (1962) 2370.
15 R. FOSTER, D. L. HAMICK AND P. J. PLACITO, *J. Chem. Soc.*, (1956) 3881.
16 *Cf.* Ref. 11, p. 137.
17 M. E. PEOVER, *Trans. Faraday Soc.*, 58 (1962) 1656.
18 G. CILENTO AND M. BERENHOLK, *J. Am. Chem. Soc.*, 84 (1962) 3968.
19 J. M. HOLLAS AND L. GOODMAN, *J. Chem. Phys.*, 43 (1965) 760.
20 *Cf.* Ref. 1, p. 47.
21 E. M. VOIGHT, *J. Am. Chem. Soc.*, 86 (1964) 3611.
22 K. NAKAMOTO, *J. Am. Chem. Soc.*, 74 (1952) 1739.
23 B. G. ANEX AND L. J. PARKHURST, *J. Am. Chem. Soc.*, 85 (1963) 3301.
23a D. J. ILTEN AND K. SAUER, *Bio-Organic Chemistry Group Quarterly Report*, UCRL-10634, January 1963, p. 69.
24 M. E. BROWNE, A. OTTENBERG AND R. L. BRANDON, *J. Chem. Phys.*, 41 (1964) 3265.
25 J. W. EASTMAN, G. ENGELSMA AND M. CALVIN, *J. Am. Chem. Soc.*, 84 (1962) 1339; I. ISENBERG AND S. L. BAIRD JR., *J. Am. Chem. Soc.*, 84 (1962) 3803.
26 B. A. BOLTO AND D. E. WEISS, *Australian J. Chem.*, 15 (1962) 653.
27 D. E. FLEISHMAN AND G. TOLLIN, *Proc. Natl. Acad. Sci. (U.S.)*, 53 (1965) 237.
28 F. KAUFLER AND E. KUNZ, *Ber.*, 42 (1909) 2482, reported the first cases.
29 K. M. HARMON, S. D. ALDERMAN, K. E. BENKER, D. J. DIESTLER AND P. A. GEBAUER, *J. Am. Chem. Soc.*, 87 (1965) 1700.
30 L. R. MELBY, R. J. HARDNER, W. R. HERTLER, W. I. MAHLER, R. C. BENSON AND W. E. MOCHEL, *J. Am. Chem. Soc.*, 84 (1962) 3374.
31 R. FOSTER AND P. HANSON, *Tetrahedron*, 21 (1965) 255.
32 Z. RAPPOPORT AND A. HOROWITZ, *J. Chem. Soc.*, (1964) 1348.
33 B. C. McKUSICK, R. E. HECKERT, T. L. CAIRNS, D. D. COFFMAN AND H. F. MOWER, *J. Am. Chem. Soc.*, 80 (1958) 2806.
34 W. R. HERTLER, H. D. HARTZLER, D. S. ACKER AND R. E. BENSON, *J. Am. Chem. Soc.*, 84 (1962) 3387.
35 T. J. WALLACE, J. M. MILLER, H. PROBNER AND A. SCHRIESHEIM, *Proc. Chem. Soc.*, (1962) 384.
36 F. J. SMENTOWSKI, *J. Am. Chem. Soc.*, 85 (1963) 3036.
37 G. A. RUSSELL, E. G. JANSEN AND E. T. STROM, *J. Am. Chem. Soc.*, 86 (1964) 1807.

38 H. Scott, G. A. Miller and M. M. Labes, *Tetrahedron Letters*, (1963) 1073.
39 N. C. Yang and Y. Gaoni, *J. Am. Chem. Soc.*, 86 (1964) 5022.
40 For a review, see L. F. Fieser and M. Fieser, *Advanced Organic Chemistry*, Reinhold, New York, 1961; L. F. Fieser and M. Fieser, *Topics in Organic Chemistry*, Reinhold, New York, 1962 and 1963.
41 E. S. Gould and H. Taube, *J. Am. Chem. Soc.*, 86 (1964) 1318, and references therein.
42 E. F. Caldin, *Fast Reactions in Solution*, Wiley, New York, 1964.
43 G. Fraenkel, in *Techniques of Organic Chemistry*, Interscience, New York, 1960, p. 2801.
44 F. J. C. Rossotti and H. Rossotti, *Determination of Stability Constants and Equilibrium Constants in Solution*, McGraw Hill, New York, 1961.
45 H. L. Schläfer, *Komplexbildung in Lösung*, Springer, Berlin, 1961.
46 W. B. Person, *J. Am. Chem. Soc.*, 87 (1965) 167.
47 K. Conrow, G. D. Johnson and R. E. Bowen, *J. Am. Chem. Soc.*, 86 (1964) 1025.
48 G. D. Johnson and R. E. Bowen, *J. Am. Chem. Soc.*, 87 (1965) 1655.
49 L. E. Orgel and R. S. Mulliken, *J. Am. Chem. Soc.*, 79 (1957) 4839.
50 S. Gorter, J. N. Murrell and E. J. Rosch, *J. Chem. Soc.*, (1965) 2048.
51 J. N. Murrell, *Quart. Rev. (London)*, 15 (1961) 191.
52 C. Reichardt, *Angew. Chem. Intern. Ed. Engl.*, 4 (1965) 29.
53 K. M. C. Davis and M. C. R. Symons, *J. Chem. Soc.*, (1965) 2079.
54 E. M. Kosower, *J. Am. Chem. Soc.*, 80 (1958) 3253.
55 Y. Matsunaga, *J. Chem. Phys.*, 41 (1964) 1609.
56 J. H. Lupinski, *J. Phys. Chem.*, 67 (1963) 2725.
57 Q. H. Gibson, V. Massey and N. Atherton, *Biochem. J.*, 85 (1962) 369.
58 H. Beinert and R. H. Sands, in *Free Radicals in Biological Systems*, Academic Press, New York, 1961, considered this material a "quinhydrone-like dimer".
59 M. W. Hanna and A. L. Ashbaugh, *J. Phys. Chem.*, 68 (1964) 811.
60 J. C. Shug and R. J. Martin, *J. Phys. Chem.*, 66 (1962) 1554.
61 D. W. Larsen and A. R. Allred, *J. Am. Chem. Soc.*, 87 (1965) 1216.
62 A. Fratiello, *J. Chem. Phys.*, 41 (1964) 2204.
63 W. M. Clark, *Oxidation–Reduction Potentials of Organic Systems*, Williams and Wilkins, Baltimore, 1960 p. 229.
64 M. E. Peover, *Trans. Faraday Soc.*, 60 (1964) 417.
65 D. R. Crow and J. V. Westwood, *Quart. Rev. (London)*, 19 (1965) 57.
66 G. H. Müller, in *Techniques of Organic Chemistry*, Vol. I, Part IV, Interscience, New York, p. 3157.
67 A. Ehrenberg and H. Theorell, in M. Florkin and E. H. Stotz (Eds.), *Comprehensive Biochemistry*, Vol. 3, Elsevier, Amsterdam, 1958, p. 169.
68 S. Udenfriend, *Fluorescence Assay in Biology and Medicine*, Academic Press, New York, 1962.
69 B. L. van Duren, *Chem. Rev.*, 63 (1963) 325.
70 N. Christodouleos and S. P. McGlynn, *J. Chem. Phys.*, 40 (1964) 166.
71 Cf. M. Eigen and G. G. Hammes, *Advan. Enzymol.*, 25 (1963) 1.
72 J. H. Swinehart, *J. Am. Chem. Soc.*, 87 (1965) 904.
73 R. L. Strong, *J. Phys. Chem.*, 66 (1962) 2423.
74 C. G. Swain and L. I. Taylor, *J. Am. Chem. Soc.*, 84 (1962) 2456.
75 A. K. Colter and S. S. Wang, *J. Am. Chem. Soc.*, 85 (1963) 115.
76 A. K. Colter, S. S. Wang, G. H. Megerle and P. S. Ossip, *J. Am. Chem. Soc.*, 86 (1964) 3106.
77 A. K. Colter and L. M. Clemens, *J. Am. Chem. Soc.*, 87 (1965) 847.

78 H. R. HORTON AND D. E. KOSHLAND JR., *J. Am. Chem. Soc.*, 87 (1965) 1126.
79 S. SHIFRIN AND N. O. KAPLAN, *Advan. Enzymol.*, 22 (1960) 337.
80 H. THEORELL AND A. P. NYGAARD, *Acta Chem. Scand.*, 8 (1954) 1649.
81 H. A. HARBURY AND K. A. FOLEY, *Proc. Natl. Acad. Sci. (U.S.)*, 44 (1958) 662.
82 H. A. HARBURY, K. F. LA NOYE, P. A. LOACH AND R. M. AMICK, *Proc. Natl. Acad. Sci. (U.S.)*, 45 (1959) 1708.
83 W. KAUZMANN, *Advan. Protein Chem.*, 4 (1959) 1.
84 E. M. KOSOWER AND P. E. KLINEDINST JR., *J. Am. Chem. Soc.*, 78 (1956) 3493.
85 E. M. KOSOWER, *J. Am. Chem. Soc.*, 78 (1956) 3497.
86 H. A. BENESI AND J. H. HILDEBRAND, *J. Am. Chem. Soc.*, 71 (1949) 2703.
87 I. ISENBERG AND A. SZENT-GYORGYI, *Proc. Natl. Acad. Sci. (U.S.)*, 44 (1958) 857.
88 D. E. FLEISHMAN AND G. TOLLIN, *Biochim. Biophys. Acta*, 94 (1965) 255.
89 B. PULLMAN AND A. PULLMAN, *Proc. Natl. Acad. Sci. (U.S.)*, 44 (1958) 1197.
90 E. M. KOSOWER, in E. C. SLATER (Ed.), *Flavins and Flavoproteins*, Elsevier, Amsterdam, 1966, p. 1.
91 V. MASSEY, G. PALMER, C. H. WILLIAMS JR., B. E. P. SWABODA AND R. H. SANDS, in E. C. SLATER (Ed.), *Flavins and Flavoproteins*, Elsevier, Amsterdam, 1966, p. 133.
92 V. MASSEY AND G. PALMER, *J. Biol. Chem.*, 237 (1962) 2347.
93 H. R. MAHLER AND L. L. BRAND, in *Free Radicals in Biological Systems*, Academic Press, New York, 1961, p. 157.
94 R. L. SEARLS AND D. R. SANADI, in *Light and Life*, Johns Hopkins, Baltimore, 1961, p. 157.
95 V. MASSEY AND N. M. ATHERTON, *J. Biol. Chem.*, 237 (1962) 2965.
96 I. ISENBERG, A. SZENT-GYORGYI AND S. L. BAIRD JR., *Proc. Natl. Acad. Sci., (U.S.)*, 46 (1960) 1307.
97 D. E. FLEISHMAN AND G. TOLLIN, *Proc. Natl. Acad. Sci. (U.S.)*, 53 (1965) 38.
98 L. D. WRIGHT AND D. B. MCCORMICK, *Experientia*, 20 (1964) 501.
99 F. S. EBERTS JR., G. SLOMP AND J. L. JOHNSON, *Arch. Biochem. Biophys.*, 95 (1961) 305.
100 J. C. M. TSIBRIS, D. B. MCCORMICK AND L. D. WRIGHT, *Biochemistry*, 4 (1964) 504.
100a J. O. HIRSHFELDER, in L. PAULING AND H. ITANO (Eds.), *Molecular Structure and Biological Specificity*, American Institute of Biological Sciences, Washington, 1957, p. 86.
101 G. KARREMAN, *Bull. Math. Biophys.*, 23 (1961) 135; *Ann. N.Y. Acad. Sci.*, 96 (1962) 1029.
102 G. CILENTO AND P. GIUSTI, *J. Am. Chem. Soc.*, 81 (1959) 3801; *J. Biol. Chem.*, 236 (1961) 907.
103 S. SHIFRIN, *Biochim. Biophys. Acta*, 81 (1962) 205; *Biochemistry*, 3 (1964) 829; *Biochim. Biophys. Acta*, 96 (1965) 173.
103a R. W. TAFT, in M. NEWMAN (Ed.), *Steric Effects in Organic Chemistry*, Wiley, New York, 1956, p. 556, for sigma values.
104 G. WEBER, *Biochem. J.*, 47 (1950) 114.
105 K. YAGI AND Y. MATSUOKA, *Biochem. Z.*, 328 (1956) 138.
106 K. YAGI, T. OZAWA AND K. OKADA, *Biochim. Biophys. Acta*, 35 (1959) 102.
106a H. BRIEGLEB AND H. DELLE, *Z. Electrochem.*, 64 (1960) 347.
107 K. YAGI, J. OKUDA, T. OZAWA AND K. OKADA, *Biochem. Z.*, 328 (1957) 492.
108 M. A. SLIFKIN, *Nature*, 193 (1962) 464.
109 I. ISENBERG AND A. SZENT-GYORGYI, *Proc. Natl. Acad. Sci. (U.S.)*, 45 (1959) 1229.
110 G. K. RADDA AND M. CALVIN, *Biochemistry*, 3 (1964) 384.
111 G. WEBER, *Nature*, 180 (1957) 1409.
112 G. WEBER, *J. Chim. Phys.*, 55 (1958) 878.

113 J. M. SIEGEL, G. A. MONTGOMERY AND R. M. BOCK, *Arch. Biochem. Biophys.*, 82 (1959) 288.
114 R. BEUKERS AND A. SZENT-GYORGYI, *Rec. Trav. Chim.*, 81 (1962) 255.
115 J. DUCHESNE, P. MACHMER AND M. READ, *Compt. Rend.*, 260 (1965) 2081.
116 P. MACHMER AND J. DUCHESNE, *Compt. Rend.*, 260 (1965) 3775.
117 E. FUJIMORI, *Proc. Natl. Acad. Sci. (U.S.)*, 45 (1959) 133.
118 M. A. SLIFKIN, *Nature*, 197 (1963) 275.
119 M. A. SLIFKIN, *Nature*, 197 (1963) 42.
120 A. SZENT-GYORGYI, I. ISENBERG AND S. L. BAIRD JR., *Proc. Natl. Acad. Sci. (U.S.)*, 46 (1960) 1444.
121 J. DUCHESNE AND P. MACHMER, *Compt. Rend.*, 260 (1965) 4279.
122 A. SZENT-GYORGYI, I. ISENBERG AND J. MCLAUGHLIN, *Proc. Natl. Acad. Sci. (U.S.)*, 47 (1961) 1089.
123 S. LIAO AND H. G. WILLIAMS-ASHMAN, *Biochem. Pharmacol.*, 6 (1961) 53.
124 M. WILK, *Biochem. Z.*, 333 (1960) 166.
125 S. G. A. ALIVASITOS, F. UNGAR, A. JIBRIL AND G. A. MOURKIDES, *Biochim. Biophys. Acta*, 51 (1961) 361.
126 D. E. LASKOWSKI, *Anal. Chem.*, 32 (1960) 1171.
127 A. PULLMAN, *Biopolymers Symp.*, 1 (1964) 49.
128 N. BROCK, H. DRUCKERY AND H. HAMPERL, *Arch. Exptl. Pathol. Pharmakol.*, 189 (1938) 709.
129 H. WEIL-MALHERBE, *Biochem. J.*, 40 (1946) 351.
130 E. BOYLAND AND B. GREEN, *Brit. J. Cancer*, 16 (1962) 347.
131 P. DESANTIS, E. GIGLIO, A. M. LIQUORI AND A. RIPAMONTI, *Nature*, 191 (1961) 900.
131a E. BOYLAND, B. GREEN AND S-L. LIU, *Biochim. Biophys. Acta*, 87 (1964) 653.
132 B. PULLMAN, P. CLAVERIE AND J. CAILLET, *Science*, 147 (1965) 1305.
133 L. PAULING AND E. B. WILSON JR., *Introduction to Quantum Mechanics*, Wiley, New York, 1935, p. 387.
134 A. L. STONE AND D. F. BRADLEY, *J. Am. Chem. Soc.*, 83 (1961) 3627.
135 L. S. LERMAN, *J. Cellular Comp. Physiol.*, Suppl. 1 (1964) 1.
136 L. S. LERMAN, *J. Mol. Biol.*, 10 (1964) 367.
137 S. H. SNYDER AND C. R. MERRIL, *Proc. Natl. Acad. Sci. (U.S.)*, 54 (1965) 258.
138 G. KARREMAN, I. ISENBERG AND A. SZENT-GYORGYI, *Science*, 130 (1959) 1191.
139 B. PULLMAN, *Biopolymers Symp.*, 1 (1964) 152.
140 S. LIAO, J. T. DULANEY AND H. G. WILLIAMS-ASHMAN, *J. Biol. Chem.*, 237 (1962) 2981.
141 A. C. ALLISON AND T. NASH, *Nature*, 197 (1963) 762.
142 L. E. ORGEL, *Quart. Rev. (London)*, 8 (1954) 422.
143 C. K. JØRGENSON, *Absorption Spectra and Chemical Bonding in Complexes*, Pergamon, London, 1962, Chapter 9.
144 P. N. MOORTHY AND J. J. WEISS, *J. Chem. Phys.*, 42 (1965) 3121.
145 C. N. SCHRAUZER AND H. THYRET, *Theoret. Chim. Acta*, 1 (1963) 172.
146 H. W. MOORE AND K. FOLKERS, *J. Am. Chem. Soc.*, 86 (1964) 3393.
147 L. E. ORGEL, *Biochemical Society Symposium No. 15*, Cambridge University Press, Cambridge, 1958, p. 19.
148 R. VERCAUTEREN AND R. MOSSART, in O. HAYAISHI (Ed.), *Oxygenases*, Academic Press, New York, 1962, p. 370.
149 F. J. DUNN AND C. R. DAWSON, *J. Biol. Chem.*, 189 (1951) 485.
150 P. O'D. OFFENHARTZ, *J. Chem. Phys.*, 42 (1965) 3566.
151 H. R. MAHLER, A. S. FAIRHURST AND B. MACKLER, *J. Am. Chem. Soc.*, 77 (1955) 1519.

152 T. KAMEDA, *Osaka Daigaku Igaku Zasshi*, 10 (1958) 29; *Chem. Abstr.*, 52 (1958) 5296g.
153 P. BAMBERG AND P. HEMMERICH, *Helv. Chim. Acta*, 44 (1961) 1001.
154 P. HEMMERICH AND J. SPENCE, in E. C. SLATER (Ed.), *Flavins and Flavoproteins*, Elsevier, Amsterdam, 1966, p. 82.
155 P. HEMMERICH, *14th Mosbacher Colloquium der Gesellschaft für physiologische Chemie*, Springer, Heidelberg, 1964.

Chapter III

Charge Transfer in Biology*

Section b

Transfer of Charge in the Organic Solid State

F. J. BULLOCK**

Laboratory of Chemical Biodynamics, University of California, Berkeley, Calif. (U.S.A.)

Processes involving transfer of charge through condensed organic matrices are not only of intrinsic interest but may also have an important bearing on many areas of biochemistry. Twenty years have now passed since Szent-Gyorgyi[1] proposed that the solid state might play a role in biological processes. The flurry of research and speculation on the role of semiconduction in biology which this suggestion eventually precipitated has already been reviewed, in whole or in part, several times[2-5]. Very recently a critical treatment of conductivity in organic crystals, proteins and donor–acceptor complexes, including some of the biological aspects, has appeared[6]. This chapter also has a good historical perspective. The same volume[7] contains a comprehensive review of the organic polymers which conduct electricity, presenting well over two-hundred references to the original literature.

Our plan in this second part will be (1) to summarize the current situation regarding transfer of charge in biological molecules *via* semiconduction; (2) to consider some of the recent results in the organic semiconductor field

* The preparation of this chapter was sponsored, in part, by the U.S. Atomic Energy Commission.
** Present address: Arthur D. Little Inc., Cambridge, Mass. (U.S.A.), NIH Postdoctoral fellow, 1964–65.

References p. 164

[149]

which seem of interest to us; and (*3*) to consider the photochemical processes which can result in generation of charge carriers and transfer of charge, an area currently yielding some very interesting results.

1. Transfer of charge by semiconduction

In this section, electrical conductivity is considered in the context of a charge-transfer process. Semiconductors are somewhat arbitrarily defined as materials having electrical properties somewhere between those of conductors and insulators. Their conductivity varies exponentially with temperature according to the relationship

$$\sigma = \sigma_0 e^{-E/kT}$$

where σ is the conductivity and E is best considered as the activation energy for conduction. Conductivity in the organics has often been considered in terms of an excitation of a valence electron to a conduction band, a picture taken over from the inorganics. Whether it is at all applicable to organic materials or whether the activation energy is in truth related to a band gap is a matter of current debate among the experts. This is a highly non-trivial topic and in the ensuing discussion we will restrict ourselves to phenomenology.

For dry proteins and nucleic acids the activation energy for conduction is large and the conductivity is accordingly very low. Hydrated proteins and nucleic acids are better conductors, but the increased conductivity is thought to be due to protons from the water which act as charge carriers. Rosenberg[8], however, feels he has shown that proteins can show electronic conduction which is intrinsic. With 7.5% water adsorbed on protein, no electrolysis of the water occurs during the conductivity experiments. This is felt to be incompatible with a conduction mechanism involving water protons.

Oxygen adsorbed on the surface of the conducting material can increase surface conduction, presumably by trapping electrons as $O_2{}^-$, facilitating conduction by the resultant positive "holes". A recent study[9] shows that conduction in purines and pyrimidines is increased by adsorbed oxygen. The conductivities are still low, however.

Direct transmission of electrons through the π-systems of proteins or nucleic acids now seems a rather unlikely possibility in living systems. Propagation of electronic excitation in biopolymers will be considered in another chapter of this volume. Recalling the phenomenon of bioluminatescence which involves generation of electronic excitation from stored chemical energy, one realizes the physical feasibility of such a process in a living system.

The highest conductivities thus far observed in organic molecules are found among the donor–acceptor complexes. The activation energies for conduction are small, on the order of 0.1 eV (~ 2 kcal) or less. Examples are the complexes of aromatic hydrocarbons with iodine and complexes of amines with chloranil or related quinones. The best conductors have a high free-radical content and are usually not of simple stoichiometry. The complex salts of TCNQ[10] (I) are the best of the known organic conductors. The

(I)

cation may be a metal ion such as cesium, the quinolinium ion, or *N*-alkylated poly-2-vinylpyridine[11]. Magnetic susceptibility studies[12] indicate that the odd electrons, which are presumably the charge carriers, are degenerate, the same situation one finds in a metal. These salts are considered to have a superposition of TCNQ molecules with the free electrons on alternate molecules (II).

(II)

A semiconduction mechanism has, on occasion been considered as a possibility for electron transfer through the cytochrome system. It is interesting that the charge-carrier mobility in copper phthalocyanine (III), an often used model for metal porphyrins, is, in fact, among the highest (about 400 $cm^2/V \cdot sec$ at $400 \,^\circ K$) thus far observed in organic materials[13]. Introduction of the copper increases the carrier mobility by two orders of magnitude over that observed in phthalocyanine itself. This has been attributed to a metal interaction which permits the molecular orbitals to be delocalized over more than one molecule. There is support for this suggestion in an ESR study of copper phthalocyanine where a metal–metal interaction can be demonstrated[14]. Other metal ion bridging groups could presumably function in a

(III)

similar way in a variety of oxidation–reduction processes. Green[15] has described mitochondrial electron transport in terms of a "biological transducer".

It is worth recalling here, in connection with models involving porphyrin-like molecules, the interesting system reported several years ago by Wang and Brinigar[16]. They prepared a polymeric material (IV) which could coordinate with oxygen at a terminal iron atom. Salt formation with poly-L-lysine permitted aqueous solutions of polymer to be prepared. The rate of oxidation of cytochrome c by molecular oxygen was enhanced ten-fold by the presence of polymer. When the unconjugated dipyridyl ethane (V) was used as a bridging group between the iron-porphyrin rings, the polymer was catalytically inert. We are not aware that the conductivity properties of this polymer have ever been studied, but one wonders whether a high electron mobility through such a system would be responsible for the observed catalysis.

Dewar and Talati[17] have reported some polymers containing chelated iron which have reasonably good conducting properties. They point out that for metal atoms which can form π-bonds with adjacent ligands using the same

(IV)

(V)

d orbitals (Fig. 1), a through conjugation is possible in the polymer. This is achieved for ligands which are coplanar and occupy sites *trans* to one another in an octahedral or square-plane complex. Their studies have thus far centered on polymers of the type (VI), where there is a through conjugation between molecules *via* the metal.

The model for the active site of *C. pasteurianum* ferredoxin, which has been proposed by Blomstrom *et al.*[18] (VII), could, if correct, have a through conjugation between iron atoms *via* the sulfur orbitals of cysteine. Ferredoxin is an electron-transport protein which participates in photoreduction reactions catalyzed by chloroplasts.

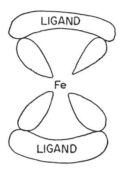

Fig. 1. Orbitals of a square-planar complex formed by two conjugated chelating ligands.

(VI)

(VII)

2. Photoconductivity

The photoprocesses involved in photosynthesis and vision have been capriciously referred to by some as "dark areas". Insofar as our understanding is concerned, this is still essentially true. When the electron microscope began revealing a number of years ago the highly organized lamellar structures in chloroplasts and the layered organization of rhodopsin in the rods of the eye it became apparent that new concepts and techniques would be required before an understanding of the function of such structures would be reached. It is very fortunate that solid-state phenomena are of such great concern to many whose major interests are outside the sphere of biology. These workers are continuing to provide both techniques and insights.

Of major concern to us here is the means by which charge carriers may be produced in organic solids by light which is energetically insufficien† to cause a direct ionization. These will be the most significant for biological photo-processes. It will be useful to first consider some of the spectroscopic consequences of forming molecular aggregates.

For crystals, one cannot consider a single molecule as the light-absorbing unit. Because the molecules are only a few Angstroms apart and the Van der Waals' interactions between molecules non-negligible, the absorption of light leads to an energy level characteristic of the assembly as a whole. The excitation is delocalized over many molecules in the crystal. This delocalized excited state, produced by a sort of excited-state resonance, is called an *exciton*. One "resonance state" in which the excitation is localized on a single molecule is pictured in Fig. 2. In some circumstances the exciton is

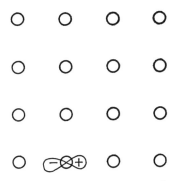

Fig. 2. Excitation localized on a single molecule of an array.

best considered as a "particle" of excitation energy hopping in a random way from molecule to molecule; in other circumstances it must be considered as a wave of excitation energy propagating through the solid.

The interaction between the transition dipoles of excited molecules in the crystal splits the original energy level into a band of levels. This is the *exciton band*. Strong interactions produce a greater splitting and lead to wider exciton bands. The lowest exciton level may be lowered with respect to the excited state of the isolated molecule. As a result, the lowest energy absorption band of the crystal may be considerably red-shifted relative to the isolated molecule. The selection rules for transitions into the exciton levels depend on the relative orientation of the transition dipole moments in the assembly. Kasha has discussed the selection rules using simple vector diagrams[19].

By a simple extension of the above ideas, it is evident that triplet excitation in crystals leads to triplet excitons and triplet-exciton bands. The energy-level diagram for crystals is then that given in Fig. 3.

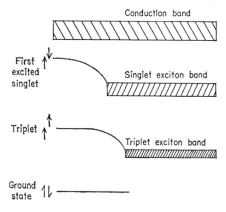

Fig. 3. Crystal-energy levels. Arrows indicate orientations of electron spin in excited and ground states.

An alternative to the tightly-bound molecular exciton pictured in Fig. 2 is a more loosely bound exciton in which the electron resides on an adjacent or nearby molecule (Fig. 4). This bound electron–hole pair is variously referred to in the literature as a Wannier exciton, a charge-transfer exciton, or an ionic exciton. It moves through the crystal as a unit and is not a charge-carrying state. The energy of this exciton state is given by

$$E_{CT} = I - EA + C + P$$

I is the ionization potential and EA the electron affinity of the molecule; C is the coulomb interaction between electron and hole, and P is the interaction (mainly dipole–dipole) due to polarizations of the lattice. This state is quite near a conducting state. It differs mainly by the energy needed to overcome the coulombic forces between the electron–hole pair, allowing them to move independently. It is therefore of interest to consider it here.

The ionic-exciton concept, introduced first by Wannier[20], has been most

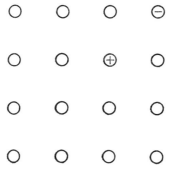

Fig. 4. An ionic exciton.

successfully used in the inorganic semiconductor field. The first application to organic crystals is apparently that of Lyons[21], but it has also been considered by Merrifield[22]. More recently the effects of configuration interaction (mixing) between ionic and neutral excitons has been considered in some detail by Jortner, Rice and co-workers[23,24], for organic crystals, and by Azumi and McGlynn[25,26], for excited dimers. The results of the crystal work are mainly of interest here.

The theoretical prediction, for aromatic hydrocarbons (the only ones now available) are as follows[23,24,27]. Ionic-exciton states will contribute mainly to neutral excitons of small band width; that is, singlet excitons which correspond to weak transitions and triplet excitons. Ionic contributions may broaden the exciton-band widths and may be particularly important in enhancing the rates of triplet-energy migration. A spectroscopic study of anthracene[28] indicates the ionic exciton lies energetically above the first singlet level. No undisputed direct experimental detection of an ionic-exciton level in a one-component organic crystal has yet been made.

The principal macroscopic effect of exciton dissociation is photoconductivity. There are several pathways by which an exciton may become dissociated to yield a conducting state. One is by interaction with a quantum of lattice vibrational energy. The vibrational energy of the lattice and the kinetic energy of the exciton supply the dissociation energy. Another pathway may be by interaction of the exciton with a defect or impurity in the crystal. Provided sufficient energy becomes available at these sites, dissociation will occur. An exciton–crystal surface interaction has also been observed to lead to conducting states. On energy-conservation grounds, a static crystal imperfection cannot readily dissociate an exciton which lies below all conducting states.

The recent discoveries of exciton–exciton interactions leading to charge carriers in the bulk of crystals are the most significant recent advances in this field, and we will present them in somewhat greater detail. The significance lies in the fact that the interaction of two excitons, a process energetically capable of yielding an equivalent of twice the excitation energy, can provide ample energy for reaching a conducting state. The advent of the laser, in particular, has signaled many of these recent advances.

All the studies of exciton interactions which we will discuss have been performed on anthracene which is, in effect, the *E. coli* of workers in the field. Choi and Rice[29] attempted to explain the fact that the response of photoconductivity with wavelength of exciting light generally parallels the absorption spectrum by invoking interaction of two singlet excitons. They demonstrated theoretically that two singlet excitons could lead to a pair of charge carriers and an unexcited molecule with adequate efficiency to account for photoconductivity. At about the same time[30] it was reported that for weakly absorbed light (4150–4550 Å) the photoconductivity of anthracene varied as the square of incident light intensity, offering support for a singlet biexciton process. Hassegawa and Schneider[31] then reported that photoconductivity could be induced in anthracene with the red (6943 Å) light from a ruby laser. Almost simultaneously it was observed[32] that the ruby-laser light (an approximately 40-kcal quantum of light) could elicit a delayed (milliseconds) fluorescence in anthracene. The characteristic blue fluorescence of anthracene corresponds to the emission of a 79-kcal photon. It was proposed that the triplet state of anthracene had been directly populated by the intense laser light. Interaction of two triplet excitons then in a (comparatively) slow process yielded an excited singlet from which fluorescence occurred. The direct population of a triplet level from the ground state is of course the reverse of a phosphorescence. The intense light beam partially compensates for the intrinsically low probability of such a transition. By studying the intensity of the delayed fluorescence as a function of exciting light in the region 5000–7000 Å they indirectly determined the spectrum of the ground-state singlet to first excited triplet transition[33] (Fig. 5). Quite recently it has been reported[34] that the photoresponse of conduction in anthracene in the 5000–7000 Å region parallels quite well (Fig. 6) this singlet–triplet spectrum. Equally interesting, a study of the quantum efficiency of the production of charge carriers revealed that the triplet exciton yields charge carriers 40 times more efficiently than the singlet exciton.

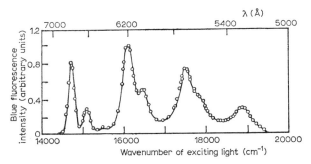

Fig. 5. Triplet-excitation spectrum of single-crystal anthracene as measured by observing the intensity of the blue fluorescence as a function of the wavelength of exciting light. (From ref. 33)

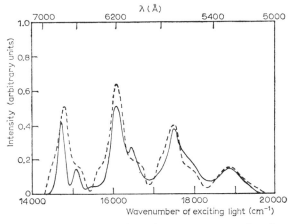

Fig. 6. Photocurrent and the triplet-absorption spectrum as a function of the exciting wavelength. ----, photocurrent; ———, blue fluorescence. (From ref. 34)

Although the exact details are not known at present, evidence is available, then, that the triplet level in addition to the singlet level can yield charge carriers efficiently. In a recent detailed theoretical consideration of triplet excitons in aromatic molecular crystals, Jortner, Rice and Katz[27] have again pointed out that the most favored triplet–triplet annihilation process should lead to a charge-transfer state. For indirect evidence in support of this contention they refer

$$2\,T \rightarrow G^+ G^-$$

to a flash spectroscopic study of Lindquist's[35]. Lindquist reports that the kinetics of decay of triplet fluorescein in solution cannot be understood unless the decay mechanism includes a reaction between two fluorescein triplets to yield a molecule each of oxidized and reduced fluorescein.

We should note here in passing that a detailed study of ruby-laser generation of excitons in anthracene crystals[36] indicates a somewhat more complicated picture than we presented above. Using a giant pulse ruby laser (capable of delivering up to 10 megawatts of power in a single 30-nanosecond —10^{-9} sec—pulse) two-photon direct excitation to a vibrationally excited singlet-exciton level was demonstrated to occur, as well as a singlet to two-triplet process (the reverse of a triplet–triplet annihilation).

Photoconductivity in copper phthalocyanine (III) now has been observed to occur with near-infrared light[37]. Weak absorption of light in the near infrared by copper phthalocyanine has previously been observed[38]. Although the ratio of extinction coefficients for visible and infrared light is 100 : 1, the ratio of the photoresponse peak heights (Fig. 7) is 3 :1. This suggests that the

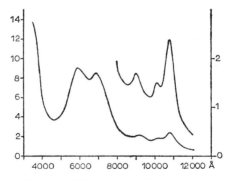

Fig. 7. Photoresponse of copper phthalocyanine. Photocurrent units are arbitrary; left scale refers to lower curve, which is the absorption spectrum. (From ref. 37)

quantum-yield ratio is about 30 : 1 in favor of the long-wavelength absorption peak. This is similar to the 40 : 1 ratio observed previously for charge-carrier production from triplet and singlet levels in anthracene[34]. The implication that the near-infrared absorption of copper phthalocyanine is a singlet to triplet transition has not been confirmed at present.

3. Donor–acceptor complexes

Since the charge-transfer transition may be considered as a first step in the production of charge carriers, it is natural that the photoconductivity of donor–acceptor complexes be investigated. The conductivity of some complexes which are relatively good conductors in the dark and show strong ESR signals can indeed be enhanced by light. We focus here, however, on recent attempts to produce by light improved conductivity in low conductivity complexes. These complexes show no ESR signal or at best only a very weak signal. Examples are the complexes of anthracene and trinitrobenzene[39], or the pyrene–TCNE complex[40]. Photoconductivity in this class of low-conductivity complexes has been inadequately investigated, but it is evident from the cases thus far reported that the photoconductivities of the solids are low. The study of Akamoto and Kuroda[40] offers some interesting features. The spectral dependence of photoconduction for several complexes did not follow the absorption spectrum (Fig. 8) and excitation in the charge-transfer

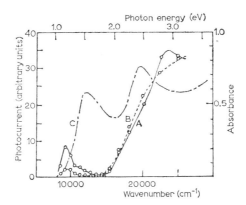

Fig. 8. Spectral dependence of photoconduction in pyrene–TCNE complex. A, sandwich cell; B, surface cell; C, absorption spectrum. (From ref. 40)

band elicited no photoresponse. This is true for both conduction in the bulk of the crystal and along its surface which can be studied separately by different electrode arrangements. The weak photocurrent peak elicited by the low-energy light showed a different temperature dependence from the main response peak and was considered to be secondary to the main peak. It was proposed that the secondary photocurrent resulted from excitation to a charge-transfer exciton state which was dissociated at a crystal imperfection.

We note however that a study of energy transfer in crystalline donor–acceptor complexes of several aromatic hydrocarbons with trinitrobenzene[41] indicates no transport of excitation energy and a rapid (10^{-8} sec) relaxation of the excited state to the ground state. This may, in fact, be responsible for the lack of photoresponse in the charge-transfer band in solid donor–acceptor complexes.

By way of contrast, the donor–acceptor complex of tetrahydrofurane and TCNE in solution shows a reversible photoresponse and ESR signal on irradiation in the charge-transfer band[42]. This is attributed to a reversible electron transfer.

4. Models of lamellar systems

Calvin has already reviewed this laboratory's interest in such systems[43,44]. In essence, it is as follows: In working from the dark reactions involving the path of carbon in photosynthesis back toward the primary light reaction, attention next fixed on the process by which electromagnetic energy is stored as chemical energy. This is the so-called quantum-conversion step. It involves, on the one hand, a chemical reduction, the conversion of pyridine nucleotide to reduced pyridine nucleotide, and on the other the oxidation of water to oxygen. The process is a very complicated one, involving two different pigment systems; but it was felt, more or less intuitively, that whatever the final details proved to be, a charge-separation step would be necessary. This was suggested by the fact that nearly all the energy of a 40-kcal quantum of light is stored as chemical energy. Under these conditions, the energy barrier preventing the products from going back to starting materials is so small that a physical separation of the oxidizing and reducing sites is necessary. Ionization of a donor–acceptor complex, involving, for example, a quinone at a trapping site, could lead to a charge separation.

Indeed, it proved[45] that evaporating a layer of o-chloranil onto a film of phthalocyanine enhanced the dark conductivity of the phthalocyanine by a factor of 10^7 and the photoconductivity by 10^5. Similar observations were made using other hydrocarbons[46].

In a somewhat similarly conceived experiment[47], carried out by painting the components (aromatic amines and various dyes) onto opposite electrodes and then clamping them together, a photovoltage of up to 3 V in room light is claimed. This is attributed to a dye-induced photo-oxidation of the amine as a first step.

5. Dye-sensitized photoconductivity

We cannot really do justice to this exceedingly interesting topic here. The technical applications of sensitized photoconductivity in photocopying methods and photographic processes are tremendous. We will refer to only one recent report as an example of this phenomenon[48]. Coating copper phenylacetylenide (VIII) powder with a dye, chlorophyll is a particularly good one, by dipping the powder in an ethanolic solution of the dye gives the material a good conductivity photoresponse with visible light. Chlorophyll itself shows a very low photocurrent[49].

$$\text{Cu}-\text{C}\equiv\text{C}-\langle\!\!\!\bigcirc\!\!\!\rangle$$

(VIII)

The above result is one more manifestation of the remarkable effects which can be elicited from a suitably contrived inhomogeneous solid. Needless to say, Nature's contrivances are even more remarkable, although she has worked at it a little longer.

REFERENCES

1   A. SZENT-GYORGYI, *Nature*, 157 (1946) 875.
2   D. D. ELEY, *Research, (London)*, 12 (1959) 293.
3   J. GERGELEY, in *Symposium of Electrical Conductivity in Organic Solids*, Interscience, New York, 1961, p. 369.
4   D. D. ELEY, in *Horizons in Biochemistry, A. Szent-Gyorgyi Dedicatory Volume*, Academic Press, New York, 1962, p. 341.
5   F. GUTMANN AND A. NETCHEY, *Rev. Pure Appl. Chem.*, 12 (1962) 2.
6   J. KOMMANDEUR, in *Physics and Chemistry of the Organic Solid State*, Vol. 2, Interscience, New York, 1965, p. 1.
7   D. E. WEISS AND B. A. BOLTO, in *Physics and Chemistry of the Organic Solid State*, Vol. 2, Interscience, New York, 1965, p. 67.
8   B. ROSENBERG, *Nature*, 193 (1962) 364.
9   S. BASU AND W. J. MOORE, *J. Phys. Chem.*, 67 (1963) 1563.
10   W. J. SIEMONS, P. E. BIERSTEDT AND R. G. KEPLER, *J. Chem. Phys.*, 39 (1963) 3523.
11   J. H. LUPINSKI AND K. D. KOPPLE, *Science*, 146 (1964) 1038.
12   R. G. KEPLER, *J. Chem. Phys.*, 39 (1963) 3528.
13   G. H. HEILMEIER AND S. E. HARRISON, *Phys. Rev.*, 132 (1963) 2010.
14   S. E. HARRISON AND J. M. ASSOUR, *J. Chem. Phys.*, 40 (1964) 365.
15   D. E. GREEN AND S. FLEISHER, in *Horizons in Biochemistry, A. Szent-Gyorgyi Dedicatory Volume*, Academic Press, New York, 1962, p. 381.
16   J. H. WANG AND W. S. BRINIGAR, *Proc. Natl. Acad. Sci., (U.S.)*, 46 (1960) 958.
17   M. J. S. DEWAR AND A. M. TALATI, *J. Am. Chem. Soc.*, 86 (1964) 1592.
18   D. C. BLOMSTROM, E. KNIGHT JR., W. D. PHILLIPS AND J. E. WEIHER, *Proc. Natl. Acad. Sci., (U.S.)*, 51 (1964) 1085.
19   M. KASHA, *Radiation Res.*, 20 (1963) 55.
20   G. H. WANNIER, *Phys. Rev.*, 52 (1937) 191.
21   L. E. LYONS, *J. Chem. Soc.*, (1957) 5001.
22   R. E. MERRIFIELD, *J. Chem. Phys.*, 34 (1961) 1835.
23   S. CHOI, J. JORTNER, S. A. RICE AND R. SILBEY, *J. Chem. Phys.*, 41 (1964) 3294.
24   R. SILBY, J. JORTNER AND S. A. RICE, *J. Chem. Phys.*, 42 (1965) 1515.
25   T. AZUMI AND S. P. MCGLYNN, *J. Chem. Phys.*, 41 (1964) 3131.
26   T. AZUMI AND S. P. MCGLYNN, *J. Chem. Phys.*, 42 (1965) 1675.
27   J. JORTNER, S. A. RICE AND J. L. KATZ, *J. Chem. Phys.*, 42 (1965) 309.
28   R. S. BERRY, J. JORTNER, J. C. MACKIE, E. S. PYSH AND S. A. RICE, *J. Chem. Phys.*, 42 (1965) 1535.
29   S. CHOI AND S. A. RICE, *J. Chem. Phys.*, 38 (1963) 366.
30   M. SILVER, D. OLNESS, M. SWICORD AND R. E. JARNAGIN, *Phys. Rev. Letters*, 10 (1963) 12.
31   K. HASEGAWA AND W. G. SCHNEIDER, *J. Chem. Phys.*, 40 (1963) 2533.
32   R. G. KEPLER, J. C. CARIS, P. AVAKIAN AND E. BRAMSON, *Phys. Rev. Letters*, 10 (1963) 400.
33   P. AVAKIAN, E. ABRAMSON, R. G. KEPLER AND J. C. CARIS, *J. Chem. Phys.*, 39 (1963) 1127.
34   R. G. SHARP AND W. G. SCHNEIDER, *J. Chem. Phys.*, 41 (1964) 3657.
35   L. LINDQUIST, *Arkiv Kemi*, 16 (1961) 79.
36   S. SINGH, W. J. JONES, W. SIEBRAND, B. P. STOICHEFF AND W. G. SCHNEIDER, *J. Chem. Phys.*, 42 (1965) 330.
37   R. DAY AND R. J. P. WILLIAMS, *J. Chem. Phys.*, 42 (1965) 4049.
38   P. E. FIELDING AND A. G. MACKAY, *Australian J. Chem.*, 17 (1964) 750.

39 M. Kleineman, L. Azarraga and S. P. McGlynn, *J. Chem. Phys.*, 37 (1962) 1825.
40 H. Akamoto and H. Kuroda, *J. Chem. Phys.*, 39 (1963) 3364.
41 R. M. Hochstrasser, S. K. Lower and C. Reid, *J. Chem. Phys.*, 41 (1964) 1073.
42 D. F. Ilten and M. Calvin, *J. Chem. Phys.*, 42 (1965) 3760.
43 M. Calvin, *J. Theoret. Biol.*, 1 (1961) 258.
44 M. Calvin, *Advan. Catalysis*, 15 (1963) 1.
45 D. Kearns, G. Tollin and M. Calvin, *J. Chem. Phys.*, 32 (1960) 1020.
46 D. Kearns and M. Calvin, *J. Am. Chem. Soc.*, 83 (1961) 2110.
47 W. C. Needler, *J. Chem. Phys.*, 42 (1965) 2972.
48 V. Mylnikov and V. Terenin, *Mol. Phys.*, 8 (1964) 388.
49 K. J. McCree, *Biochim. Biophys. Acta*, 102 (1965) 90.

Chapter IV

Active Transport and Ion Accumulation

PETER MITCHELL

Glynn Research Laboratories, Bodmin, Cornwall (Great Britain)

1. Introduction

Up to the present time, the study of the transport of substances in living organisms has been more a matter of physiology than of biochemistry, because the relationships between the biochemical reactions of metabolism and the complex biophysical processes of active transport have proved to be particularly difficult to establish at the molecular level of dimensions. Progress in this important field of biochemistry has been hampered, not only by the inevitable experimental difficulties associated with the study of the fragile multiphase systems in which the natural transport reactions occur, but also by gratuitous conceptual difficulties that are still perpetuated in the phrase "active transport". During the last few years, however, especially since Skou[1-3] initiated the work of characterising the Na^+,K^+-translocating ATPase system of plasma membranes, some significant biochemical knowledge of transport reactions has been obtained.

In order to use the space allotted to the present chapter to the best advantage, I shall mainly concentrate attention upon those limited areas of the field of biological transport in which practical and theoretical biochemical advances have been made or are in progress.

(a) Some relationships between chemical reaction and transport

It is customary to consider the metabolic flow processes that occur in living organisms as consisting of chemical reactions on the one hand, and transport reactions on the other. The chemical reactions are defined as the processes of

References p. 192

reassortment of the units of organic chemical structure such as electrons, atoms, chemical groups and molecules; while the transport reactions are defined as the processes of diffusion from one place to another of the chemical structural units in various states of assembly. The same kind of distinction is made in most considerations of chemical synthesis or degradation in the test tube or factory. The transport of reactants up to and of resultants away from the "point" of chemical reaction is usually considered separately from the chemical change itself (see *e.g.* Bosworth[4]). This useful distinction has permitted chemistry to develop largely on the basis of a scalar terminology designed to define initial and final states in terms of chemical composition, concentration, pressure, temperature, volume, entropy, energy, etc. On the other hand, the transport processes connected with the bringing up and taking away of the chemical reactants and resultants, and with the transitional intermediates of chemical change, can only be sufficiently specified in space-time by the use of vectors or of tensors of higher order. Knowledge of transport phenomena has accordingly grown largely on the basis of a tensor terminology.

The question of the mechanism of "active" or metabolically coupled transport is, of course, concerned with the relationship between chemical reaction and transport. An interesting paradox has arisen in connection with the definition of active transport because coupling between the chemical reactions of metabolism and the processes of transport seems to require that scalar chemical processes should drive vector transport processes, thus contravening a principle attributed to Curie[5a,b], which is usually described in some such form as that "it is impossible for a force of a given tensorial order to be associated with a flow of a higher tensorial order"[6]. This matter has been discussed for some years (see refs. 7–9), and in view of the attention that it has recently received, we shall attempt some clarification of the fundamental issues.

Kedem[10] attempted to circumvent the paradox implicit in the driving of vectorial transport processes with supposedly scalar chemical forces by introducing a vectorial cross coefficient in a non-equilibrium thermodynamic definition of active transport. Jardetzky[9], however, stated that the direct coupling between a metabolic reaction and a transport process, implied by Kedem's vectorial cross coefficient, was impossible because it would contravene the Curie principle (see also ref. 11). Katchalsky and Kedem[12], later supported by Moszynski *et al.*[13], answered the criticism of Jardetzky[9] by saying that Langeland[14] had shown that the principle of Curie "applies only to isotropic media but not to an anisotropic membrane capable of active transport". The responsibility for the vectorial effect of the chemical reaction was thus placed upon unspecified anisotropic characteristics of "the medium". This would be consistent with the sophisticated discussion of "Curie's theorem" given by Fitts[15]. Jardetzky[11] objected to the ambiguity inherent in Kedem's attempted solution of the active transport paradox, and suggested that the correct solution was the one originally given by Jardetzky and Snell[6] who had concluded that "the one and only way to effect a transport process by the scalar forces of chemical reactions is to have a source and a sink set up by such reactions in distinct but contiguous regions of space".

Many of those engaged in studies of membrane transport (such as Wilbrandt and Rosenberg[16]) have come to accept the premise that the forces of chemical affinity are scalar. Much energy has been expended in attempts to give an exact and satisfactory definition of "active transport" from the experimental point of view, especially by Ussing and co-workers[17-21], but no general agreement has been reached. It has therefore been suggested[16,22] that it would be advisable to use alternative descriptions in place of the phrase "active transport". Perhaps the most confusing development in this context has been the proposal by Wilbrandt and Rosenberg[16] that "active transport" should be called "uphill transport" and the rejoinder of Bricker, Biber and Ussing[23] that "active transport" can be "active downhill transport".

A true paradox has no satisfactory solution, for it arises from the use of false or incompatible premises. The premise that the forces involved in chemical reactions are scalar is false (see refs. 24–26). The statement of Curie[5a] that is relevant to our considerations of the connection between transport and chemical reaction is "Lorsque certains effets révèlent une certaine dissymétrie, cette dissymétrie doit se trouver dans les causes qui lui ont donné naissance" — i.e. effects cannot be less symmetric than their causes. It follows that if a chemical force causes a vectorial transport, the chemical force cannot be scalar. The fact that chemical forces can be directed in space has, of course, long been known to chemists and biochemists.

In the same year (1941) as Lipmann[27] published his remarkable exposition of the principle of group potential and the concept of enzyme-catalysed group transfer, a comprehensive kinetic theory of the thermally-activated diffusion of chemical particles through the transitional intermediates of chemical reactions was published by Glasstone, Laidler and Eyring[28]. These precocious theoretical developments established the basis for what should be the present integral view of chemical reaction and transport phenomena. Yet, for practical reasons that we shall illustrate briefly, this integration of ideas is not yet well established.

In the interest of precision and simplicity, enzymologists prefer to study and to define the characteristics of enzyme-catalysed reactions in "homogeneous aqueous solution" under given conditions of temperature, pressure, pH, ionic strength, etc. etc. The reaction studied generally consists of the transfer of a given chemical group from a group donor to a group acceptor, and the main function of the enzyme is to lower the free energy of activation of the diffusion of the group undergoing transfer from the donor group to the acceptor group, so that the exchanging covalencies of the transition-state complex behave like the secondary valencies that generally control diffusion reactions. The part of the enzyme-catalysed group-transfer reaction that corresponds to the passage of the group undergoing transfer from the donor to the acceptor group does not, in reality, occur in the "homogeneous aqueous solution" in which the enzyme-catalysed reaction is ostensibly conducted. Rather, this vectorially-defined diffusion process, along the reaction coordinate of the transition-state complex, occurs in the anisotropic microscopic non-aqueous phase otherwise called the "active centre region" of the enzyme. Nevertheless, it is a convenient and simplifying fiction to regard enzymes in this type of context as "dissolved" reactants and to treat overall group-transfer reactions as occurring in homogeneous solution. In particular, it is thus possible to neglect the real microscopic vectorial characteristics of group-transfer reactions because the time integral of the distribution and orientation of the particles in aqueous enzyme and substrate solutions gives the equivalent of an isotropic continuum. Hence velocity constants can be given as scalar numbers and the state of the system can be described in the customary thermodynamic scalar terms of concentration, pressure, volume, temperature, entropy, energy, etc.

It is unfortunate that the usefulness of the fiction of homogeneity in this and other types of heterogeneous reaction system should have helped to foster the fallacy that chemical reactions are intrinsically scalar phenomena.

(b) Chemiosmotic processes

According to the classical thermodynamic treatment of chemical reactions, the forces of chemical reaction are defined by the chemical or group potentials which represent the tendencies of pairs of chemical groups to escape from each other and join other neighbouring partners (refs. 27,29–32). The overall tendency of a chemical reaction to go forward, represented by the standard free energy of the reaction ΔG^0, is given by the sum of the forward pressures minus the sum of the backward pressures at equilibrium, or

$$\Delta G^0 = (\bar{\mu}_i - \bar{\mu}_i^0)_{\text{Reactants}} - (\bar{\mu}_i - \bar{\mu}_i^0)_{\text{Resultants}} \tag{1}$$

where the $\bar{\mu}_i$ values are the total chemical potentials of the components of the reaction at equilibrium, and the $\bar{\mu}_i^0$ values represent those of a standard state. The activities (a_i) of the components are related to the $\bar{\mu}_i$ values by the relationship

$$\bar{\mu}_i = \bar{\mu}_i^0 + RT \ln a_i + \bar{v}_i P + z_i F \psi + \text{etc.} \tag{2}$$

where P is the pressure, ψ is the electric potential, $\bar{v}_i$ and z_i are the molecular volume and valency of i, and F is the faraday. As the pressure, electric, and other energy terms are not usually included explicitly in defining chemical-equilibrium constants in condensed systems, eqn. 1 is a simple and more general way of writing the familiar equation

$$\Delta G^0 = -RT \ln K \tag{3}$$

where K is the equilibrium constant. It thus follows that for the condensed systems in which biological metabolism and transport usually occur, ΔG^0 corresponds to the osmotic work done in bringing the reactants and resultants from the standard state to the equilibrium state. This is illustrated for the case of the hydrolysis of an anhydride AB in Fig. 1. The reaction $AB + H_2O \rightleftharpoons AH + BOH$ is supposed to be catalysed in a region specifically accessible to AB and H_2O from the left, and to AH and BOH from the right, *via* the substrate-specific membranes M_1 and M_2. The pistons P_1 and P_2 are respectively specifically impermeable to AB only and to AH and BOH only, and AB and AH + BOH are respectively present in the standard state ($\bar{\mu} = \bar{\mu}^0$) outside the pistons P_1 and P_2. The difference between the pressures on the pistons is proportional to $-\Delta G^0$.

The chemiosmotic system of Fig. 1 catalyses the translocation of the chemical groups A, B, H and OH and transforms the chemical work defined by ΔG^0 into the energetically equivalent osmotic potential in a form that can do external work as the reaction proceeds from left to right. Conversely, the osmotic potential would be converted to chemical "bond energy" when the reaction progressed from right to left as a result of appropriately adjusting

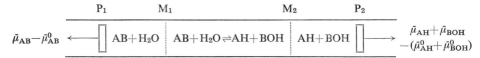

Fig. 1. Explanation in the text.

the pressures on the pistons so that the required work would be done on the system.

It might seem surprising that eqn. 1 (which equates scalar functions) should be closely related to the vectorial system of Fig. 1. It is helpful to recall, however, that the free-energy function differs from certain other scalar energy functions (such as the heat-energy function Tds, or the chemical energy function $\mu_i dn_i$) in that it is the scalar product of two co-linear vectors, such as force and displacement $(\vec{f} \cdot \vec{dx})$.

To summarise, in chemiosmotic systems the forces of chemical affinity exhibit a net vector component which may be realised macroscopically as an osmotic force. Alternatively the vector component of the forces of chemical reaction in a chemiosmotic system may not be transmitted to a macroscopic external system (as in a chemical or "heat" engine) but may result in a local flow of material. In any case, it is evident that, in accordance with the symmetry principle of Curie[5a,b], chemiosmotic reaction processes must be catalysed in such a way that the forces of chemical affinity are continuous with those of osmotic potential. Incidentally, it is interesting that all engines that transform chemical work to available free energy must be asymmetric in the sense of the present context.

It may be noted that the term osmotic has been used in a broader sense here than that defined, for example, by Guggenheim[29]. Here, the word osmotic has been used to describe processes driven by the general escaping tendency of chemical particles as they diffuse towards equilibrium. This, broader, usage of "osmotic" is closely related to the definition of the total chemical potential $(\bar{\mu}_i)$ given above, and to the definition of electrochemical activity given by Ussing[18].

2. Translocation catalysis

(a) The facilitation of solute diffusion by catalytic carriers

The main factors determining the conductance of the medium to chemical particles in biological systems have recently been nicely illustrated by the work on the conduction of oxygen through solutions of myoglobin, haemo-globin and other respiratory carriers[33-35].

The calculations of Wyman[34] indicate that under normal physiological conditions, the myoglobin present in the cytoplasm of muscle cells is respon-sible for carrying the greater part of the oxygen between the sarcolemma and the mitochondria. In the case of a monovalent carrier such as myoglobin it can be shown[36] that the rate of rectilinear diffusion under conditions of dissociation equilibrium is given by

$$\frac{dn_A}{dt} = -\frac{d\mu_A}{dx}\left\{\frac{[A]}{f_A} + \frac{[\overline{X}]}{f_X}\frac{K_A[A]}{(K_A + [A])^2}\right\} \tag{4}$$

where A stands for the solute, X stands for the carrier, μ means chemical potential, n means quantity per unit area, f stands for a frictional coefficient, t means time and square brackets denote concentration. The quantity $[\overline{X}]$ means the sum $[X]+[XA]$, and K_A stands for the dissociation coefficient of X for A.

Eqn. 4 shows that for a given gradient of chemical potential of the solute, oxygen, the carrier, myoglobin, catalyses diffusion by the factor α, given by

$$\alpha = 1 + \frac{f_A K_A [\overline{X}]}{f_X (K_A + [A])^2} \tag{5}$$

This catalysis factor is maximal if $[A] \ll K_A$, when it simplifies to

$$\alpha = 1 + \frac{f_A [\overline{X}]}{f_X K_A} \tag{6}$$

The constants given by Wyman[34] lead to:

$$f_A/f_X = 0.091 \qquad [\overline{X}] = 180 \ \mu M \qquad K_A = 5.4 \ \mu M$$

and give a value of 4 for α. These and other predictions of the translational facilitated diffusion theory given by Snell[35] and by Wyman[34], reviewed in more detail elsewhere[36], are in fair accord with the observations on the catalysis of the diffusion of oxygen by myoglobin, haemoglobin and other blood pigments.

The original definition of facilitated diffusion given by Danielli[37] related to solute translocation through non-aqueous membranes, but there is no difference in principle between facilitated diffusion in aqueous and non-aqueous media. Eqns. 4-6 are therefore of general interest as a summary of the factors affecting translocation by a circulating catalytic carrier system. The extent to which the translocation catalysis factor α exceeds unity depends upon the following factors: (a) the ratio of the frictional coefficients for the free component and carrier or carrier complex (f_A/f_{XA}); (b) the effective "solubilisation" of the component, A, in the medium through which translocation occurs by specific combination with the carrier, given by the factor $[\bar{X}]/(K_A + [A])$; (c) the force on the carrier complex relative to that on the free component, given by the factor $K_A/(K_A + [A])$.

Our discussion of translocation catalysis has so far been confined to solute translocation, but the same general principles apply to the translocation of other types of chemical particle.

(b) Group translocation

The word solute has been used to specify stable, chemically complete, particles. Solutes are secondarily bonded to the materials of the condensed phases in which they generally reside in biochemical systems; and the translocation reactions of solutes relate to the opening and closing of the secondary or ionic bonds between the solutes and the chemical components of the media through which they diffuse. However, many biochemical interactions occur between the chemically incomplete particles such as electrons, radicals, unstable ions, etc. that can conveniently be described as chemical groups. The metabolic reactions consist of the interchange between chemical groups, facilitated by the appropriately catalysed diffusion of the groups through transition-state complexes in the active-centre regions of metabolic enzymes. At the microscopic level, these specialised biological diffusion reactions differ from normal solute diffusion only in that the diffusing particles are not whole solutes, and in that the diffusion processes involve the opening and closing of primary bonds in addition to the opening and closing of secondary bonds[25].

The important suggestions of Lundegardh[38] led to the view that there might be biologically significant connections between metabolic oxido–reduction reactions, mediated by the cytochrome system, and electron translocation through natural membranes (refs. 39–43). Our object is to carry Lundegardh's suggestions to their logical conclusion by recognising the general applicability of the principles of translocation catalysis, as discussed above, to chemical groups as well as to solutes. It will be noted that, for the catalysis of the translocation of a chemical group (which must of course, include the catalysis of a chemical reaction), precisely the same factors are involved, as those described above for the case of solute translocation. Factor (a) corresponds to the lowering of the activation energy for passage through the transition-state complex relative to that for the uncatalysed reaction; (b) corresponds to the increase in concentration of the transition-state species by the presence of the catalyst; and (c) corresponds to the effective chemical potential gradient across the reaction coordinate of the transition-state complex relative to the overall free energy of the reaction.

(c) Classification of translocation reactions

It is convenient to classify translocation reactions and reaction systems as follows.

(A) In *primary translocation reactions*, primary bonds are exchanged or electrons are transferred between different pairs of chemical groups. This class of reactions may be divided into two sub-classes: (1) In *group translocation reactions*, chemical groups or electrons pass from one side of an osmotic barrier to the other[44,45]. (2) In *enzyme-linked solute translocation reactions*, the translocation of one or more solutes through the osmotic barrier is coupled to the translocation of chemical groups or electrons otherwise than across the osmotic barrier[25].

(B) In *secondary translocation reactions*, primary bond exchanges or the transfer of electrons do not occur. This class of reactions may be divided into three sub-classes: (1) In *non-coupled solute translocation*, or facilitated diffusion[37], or *uniport* reactions, a single solute equilibrates across an osmotic barrier. (2) In *sym-coupled solute translocation*, or cotransport[46], or *symport*[25] reactions, two solutes equilibrate across an osmotic barrier, and the translocation of one solute is coupled to the translocation of the other in the same direction. (3) In *anti-coupled solute translocation*, or countertransport[16], or *antiport*[25] reactions, two solutes equilibrate across an osmotic barrier, and the translocation of one solute is coupled to the translocation of the other in the opposite direction.

We shall refer to the systems catalysing secondary translocation reactions as "translocators" or "porters", avoiding the termination "ase" in conformity with the recommendations of the Commission on Enzyme Nomenclature[47]. Also, in conformity with these recommendations, the systems catalysing group translocation may be called "translocases", although, as a rule, the normal enzyme name appears to be more appropriate.

References p. 192

3. Translocation catalysis through lipoprotein membranes

Most of the substrates whose movements are controlled by the lipoprotein membranes of living tissues are very lipophobic, and have extremely low solubilities in the predominantly lipoid medium of the osmotic barrier phase of the membrane systems[48,49]. The lipid:water partition coefficients for hydrophilic solutes are generally smaller than 10^{-5}. When the translocation of a hydrophilic solute, A, through a lipoprotein membrane is catalysed by a translocator, X, the factor $[\overline{X}]/(K_A+[A])$ may therefore be large, and the catalysis factor α may have a correspondingly high value. Assuming that the concentration of solute in the aqueous media were about 1 mM, the value of [A] (concentration of A in the osmotic barrier phase of the membrane) would be less than 10^{-8} M. If the concentration of carrier in the membrane were some 200 μM, and if the carrier were about half-saturated in the membrane, the catalysis factor, α, would be of the order of 1000 or more even if f_A/f_X were as small as 0.1. Similar considerations apply to the translocation of chemical groups, and in this case the factor $[\overline{X}]/(K_A+[A])$ refers to the group concentration in the form of the transition-state complex of the translocase compared with that in the form of the solute from which the group is derived.

The non-catalysed reaction being so slow, it is often legitimate to neglect it in equations describing the catalysed reaction rate, thus

$$\frac{dn_A}{dt} = -\frac{d\mu_A}{dx}\frac{[A]}{f_X}\frac{[\overline{X}]}{(K_A+[A])}\cdot\frac{K_A}{(K_A+[A])} \tag{7}$$

Using the relationship between chemical potential and activity (represented by curly brackets)

$$\mu_A = \mu_A^0 + RT\ln\{A\} \tag{8}$$

where μ_A^0 is a standard potential, and assuming that $[A]=\{A\}$, eqn. 7 may be written

$$\frac{dn_A}{dt} = -\frac{d[A]}{dx}\cdot RT\frac{[\overline{X}]}{f_X}\frac{K_A}{(K_A+[A])^2} \tag{9}$$

Integrating over a finite concentration difference ($[A]_R-[A]_L$) between the right and left sides of the membrane of thickness x, we obtain

$$\frac{dn_A}{dt} = RT\frac{[\overline{X}]}{f_X}\frac{([A]_L-[A]_R)}{x}\cdot\frac{K_A}{(K_A+[A]_L)(K_A+[A]_R)} \tag{10}$$

Eqn. 10 may be written as

$$V = V_{max} \frac{([A]_L - [A]_R) K_A}{(K_A + [A]_L)(K_A + [A]_R)} \tag{11}$$

where V stands for dn_A/dt and V_{max} is the maximal value of V.

Eqn. 11 is identical to that given for the catalysed translocation of non-electrolytes by Wilbrandt and Rosenberg[16], and derived earlier by several authors (refs. 51–53). When $[A]_R$ is zero, eqn. 11 has the same form as that given by Michaelis and Menten[54] for enzyme kinetics

$$V = V_{max} \frac{[A]}{K_A + [A]} \tag{12}$$

In many cases it has been found that translocator-catalysed reactions obey a relationship of this kind under appropriate conditions. When $[A]_L$ and $[A]_R$ are small compared with K_A, the relationship between V and $([A]_L - [A]_R)$ is linear, as in the Fick law, or in enzyme-catalysed reactions at low saturation, or

$$V = V_{max} \frac{[A]_L - [A]_R}{K_A} \tag{13}$$

On the other hand, when $[A]_L$ and $[A]_R$ are large compared with K_A, the back reaction becomes significant, and

$$V = V_{max} \frac{([A]_L - [A]_R) K_A}{[A]_L \times [A]_R} \tag{14}$$

In this case, the translocation becomes faster as $[A]_L \times [A]_R$ is *diminished*, other things being equal. This type of "regulatory" effect is well-known in enzyme-catalysed reactions at high end-product concentration. In cases where it is not legitimate to neglect the uncatalysed reaction, a correction term corresponding to the first term on the right hand side of eqn. 4 is introduced[55,56].

Following Widdas[50], we have assumed that the carrier equilibrates relatively fast with the substrate compared with the rate of translocation, as in most enzyme reactions. When this is not so, a new rate-limiting term enters into the equations. This has been discussed in some detail by Rosenberg and Wilbrandt[53,57], Wilbrandt and Rosenberg[16], and Dawson and Widdas[58]. It does not seem to have been generally appreciated that translocator-catalysed reactions in which the carrier site does not equilibrate completely with the passenger are precisely equivalent kinetically to enzyme-catalysed reactions in which the active-centre site does not equilibrate completely with the substrate.

4. The general mechanisms of translocation catalysis

Danielli[37] pointed out that the catalysts that translocate hydrophilic solutes through the hydrophobic osmotic barrier of the lipoprotein membrane systems must permit the water that is secondarily bound to the solutes to be replaced by alternative bonding groups of approximately equivalent free energy. The necessity for this, or for some energetically equivalent compensatory process, such as that suggested by Hammes[59] for enzyme-catalysed reactions, stems from the fact that chemical particles can diffuse through the membrane phase at a significant rate only in so far as the resulting overall changes of bonding and configuration are not accompanied by free-energy changes much in excess of the thermal energy kT, corresponding to any position of the particle along the translocation pathway.

(a) Mobile versus fixed carriers

The main differences between mobile and fixed types of carriers (see refs. 33, 37, 53, 60–63) have been well summarised in the context of electron and hydrogen transport through the respiratory chain by Chance et al.[64]. They compared the current flow and fluid flow models of Holton[65] and Lundegardh[66] with the normal kinetic oxido–reduction model, and pointed out that the latter consists of a series of bimolecular reactions, while the former is equivalent to a unimolecular process. Hence, in the fixed carrier model there is effectively a single chemical channel, whereas in the bimolecular or circulating mobile carrier model there are two chemical channels.

(b) The carrier centre

Attention should be drawn to the change of our conception of the "carrier" from that modelled on a freely diffusing molecule in a homogeneous fluid (such as myoglobin in saline) to that modelled on the relatively restricted conformational changes that may be permitted in the catalytic translocation system of the membrane. The use of the word "translocation" is particularly apposite in this context (in preference to the word transport) because we mean to imply the change of accessibility of a chemical particle from one side of the osmotic barrier to the other. Little or no movement of the chemical particle itself need be involved, but the translocation must involve at least the movement of a bonding relationship (an electron) in the carrier centre of the translocator[63]. Translocation involves the *relative* movements of the substrate and the "carrier centre" of the translocator. We shall refer to the "carrier centre" rather than to the "carrier" that combines with the particle being translocated in precisely the same sense as it is customary to refer to the "active centre" of an enzyme at which the substrate is adsorbed prior to group transfer.

5. Secondary translocation

(a) Non-coupled solute translocation: uniport

(i) Circulating carrier uniporters

The best characterised of the membrane-located uniporters are the systems for D-glucose and for L-leucine translocation in the erythrocytes of mammals. LeFevre[67] has reviewed specificity and kinetic studies on the glucose uniporter of human erythrocytes which show that: (*1*) The system exhibits a rather broad substrate specificity that nevertheless indicates a three-dimensional shape relationship between the substrate and a carrier centre in the system. (*2*) The kinetics of the translocation reaction agree with a circulating carrier or bonding mechanism, such as we have outlined above, the reaction rate being limited by the translocation of the carrier centre–substrate complex within the uniporter. (*3*) Well-coupled antiport reactions can be obtained, using suitable pairs of substrates — indicating that the translocation of the substrates could not be explained by a "single chemical channel" type of mechanism. (*4*) Specific inhibitor and competition studies illustrate the close relationship between the kinetics of the uniporter-catalysed translocation and classical enzyme-catalysed group-transfer processes.

LeFevre[68,69] observed that the affinity of the glucose uniporter (represented by the apparent dissociation constant of the carrier centre, X, for the sugar, A) ranged over a factor of about 400 for the following sugars (in order of declining affinity): D-mannose, D-galactose, D-xylose, L-arabinose, D-ribose, D-arabinose. On the other hand the maximum velocity (V_{max}) of translocation for each sugar was the same, within experimental error.

The L-leucine uniporter of human and rabbit erythrocytes studied by Winter and Christensen[70,71], for instance, shows similar characteristics to the glucose uniporter. The constitutive α-thioethyl D-glucopyranoside translocation catalyst of *Saccharomyces cerevisiae*[72] is probably a typical uniporter also.

(ii) "Single channel" or "pore" uniporters

It has been known for some time that certain antibiotic polypeptides exert a rather specific effect upon natural membranes[73]. Recent observations by Pressman[74,75], however, have led to the discovery by Chappell and Crofts[76,77] that certain polypeptides of the valinomycin and gramicidin group are specific cation conductors.

Although there is still some difference of opinion, it is probable that further studies on valinomycin and gramicidins A, B, and C will confirm that they are uniport catalysts of the pore or "single chemical channel" type (ref. 36); and that whereas valinomycin has a rather high specificity for K^+ translocation and discriminates against H_3O^+ ions, owing to the effectively larger pore size, the gramicidins have a broader alkali-cation specificity and can translocate protons. The fact that the structures of valinomycin and the gramicidins are known (refs. 78–81) should encourage studies of the structure–function relationships of these important translocators.

(b) Anti-coupled solute translocation: antiport

The circulation of a carrier or of a bonding relationship between two or more channels permits a type of coupling that is not possible for the fixed or single-channel type of carrier. If there are two or more substrates that meet both the specificity requirements for combination with the carrier centre and for mobility of the carrier centre, an asymmetry of distribution of one of the substrates causes a corresponding asymmetry of distribution of the free carrier centre in the opposite direction, and thus gradients of the other substrate-carrier complexes are set up and corresponding flows occur. This exchange-diffusion type of mechanism — or antiport, as we call it — was first invoked by Ussing[17,82] to explain the coupling between the inward and outward translocation of two isotopes of Na^+ through the plasma membrane of skeletal muscle cells. An analogous system coupling the flows of two chemically different substrates that compete for the same carrier was postulated by Widdas[50] and this type of antiport has been experimentally observed in sugar[83,84], amino acid[70,85,86] and inorganic ion[87] translocation reactions.

The tightness of coupling in antiport reactions may depend both upon the degree of saturation of the carrier centre with the two substrates, and upon the mobility of the free carrier centre compared with that of the carrier centre combined with either substrate (see refs. 36, 88).

A considerable number of examples of artificial antiport reactions are now known. However, there are, as yet, no well-characterised examples of natural antiporters, although reasons have been given for thinking that such systems may exist[25,89]. Preliminary studies have indicated the existence of a system that catalyses a tightly-coupled antiport of Na^+ and H^+ across the cristae membrane of mitochondria, exhibiting a high temperature coefficient and discriminating against K^+ (ref. 43).

(c) Sym-coupled solute translocation: symport

(i) The Na^+–glucose symporter

The observations of Riklis and Quastel[90], Csaky and Thale[91], Crane et al.[92] and Smyth[93,94] on the translocation of glucose and non-fermentable sugars through the luminal border of the cells of hamster, guinea-pig and toad intestinal mucosa gave evidence for sym coupling between the translocation of Na^+ and sugar. In a remarkable research programme, Crane and his collaborators (refs. 95, 96) have gone a long way towards characterising the Na^+-glucose translocator of the brush border of the epithelium of the small intestine[97]. Extensive specificity studies[98] indicated the presence of a single carrier common to all the so-called actively transported sugars. The specificity of the overall translocation reaction and the characteristics of phlorizin inhibition[99,100] gave strong support to the view that the sugar was bound by a specific carrier site. It was further shown that the translocation of the sugar was accompanied by the translocation of Na^+, and it could be inferred that a binding site specific for Na^+ must exist on the carrier and that sugar and Na^+ moved across the membrane together.

Schultz and Zalusky[101–103] have shown by measurements of the transmural potential and short-circuit current in the rabbit ileum that a flow of electric charge (presumably carried by Na^+) accompanies the sym translocation of Na^+ and glucose or non-metabolizable glucose derivatives across the membrane system. Lyon and Crane[96] have followed up this work, using hamster small intestine, and have substantially confirmed Schultz and Zalusky's findings.

A number of studies on the specificity and kinetics of the Na^+–sugar symporter in intestinal mucosa have added weight to the general conclusions of Crane's research group outlined above (see refs. 104–110). In a recent review, Csaky[111] has again given reasons for believing[112] that "in the glucose and other Na^+-linked substrate transports the sodium dependence is connected with the "active" part of the transport, *viz.* the one that is responsible for the conversion of the chemical into osmotic energy."

Owing to observations on certain competitive and reciprocal antiport relations between sugar and amino acid transport, Alvarado and Crane[113] and Alvarado[114] have recently postulated a new "polyfunctional, mobile carrier system . . . . involved in the uphill transport of sugars, neutral amino acids and basic amino acids in the small intestine" that consists of a "mosaic of fixed, specific membrane sites which acquire mobility as a result of deformations of the "mobile membrane" resulting in local, transient engagements of the two protein surfaces, thus allowing bound substrates to be alternately exposed to the extra- and intercellular fluids."

The Na^+–glucose symporter has been studied most intensively in intestine, but similar translocation catalysts appear to occur in a number of other tissues, for example, in kidney[115]; in diaphragm[116,117]; in leucocytes[118]; and possibly in Ehrlich ascites tumour cells[119].

(ii) Na^+–amino acid symporters

The transport *in vitro* of amino acids depends upon the presence of Na^+ in the extracellular medium for Ehrlich ascites cells[120,121], brain cortex[122,123], thymus nuclei[124], intestine[125], and leucocytes[126]. Kromphardt *et al.*[127], stimulated by the work on the Na^+–glucose symporter showed that in Ehrlich ascites cells, glycine translocation is strictly Na^+-dependent. Fox *et al.*[128] made similar observations on the Na^+ dependence of transport of glycine and α-amino-isobutyric acid, and (to some extent) of lysine, in rat kidney. Christensen and Liang[129] inferred from the results of competition and kinetic studies that Ehrlich ascites cells contain six different amino acid translocators. Begin and Scholefield[130] studied a translocator for L-proline in mouse pancreas, having the kinetic properties of a divalent mobile carrier system with identical affinities for the two proline-combining sites. They suggested that there are at least three separate amino acid translocators in this tissue[131]. Blasberg and Lajtha[132] studied the substrate specificity of amino acid translocation in mouse brain. As in previous studies, several amino acid translocators were inferred, but not a specific one for each amino acid. At least six translocators were thought to be present, each one utilised by a group of amino acids of similar charge and structure. The groups were (*i*) small neutral, (*ii*) large neutral, (*iii*) small basic, (*iv*) large basic, (*v*) acidic, (*vi*) γ-aminobutyric acid. A given amino acid appeared to enter by more than one site.

It is difficult to identify the individual translocators in a multiple system having overlapping specificities, and this problem has recently been well discussed by Munck[133]. He showed by competitive-inhibition studies that in the rat small intestine there is a Na^+–imino acid symporter, specific for proline, hydroxyproline and sarcosine, that will also translocate glycine, betaine, leucine and alanine. Glycine and alanine appear to use both a neutral amino acid translocator and the imino acid translocator, confirming a suggestion of Newey and Smyth[134] that glycine enters by two routes. Munck's observations also agree with those of Oxender and Christensen[135] and Oxender[136] on the existence of two systems for neutral amino acid translocation in Ehrlich ascites cells. Their alanine-preferring system seems to correspond to the Na^+–imino acid symporter of rat intestine, while their leucine-preferring system corresponds to the neutral amino acid translocator of rat intestine. Oxender[136] has shown that these two systems are stereospecific towards the L-isomers of alanine and leucine.

At the present early stage of development of this subject there is considerable confusion as to the functional identities of given translocators. In particular, it is not yet clear in many cases, whether a given translocation reaction is directly Na^+-linked. The position is exceptional in the case of the Na^+–glycine symporter of pigeon red cells.

Vidaver[56,137–139] observed that the translocation of glycine by pigeon red cells could be analysed into two flow components, a Na^+-independent flow that was proportional to the concentration difference, and a Na^+-dependent flow, showing Michaelis–Menten kinetics. A straight line was obtained by plotting the reciprocal of the Na^+-dependent velocity of glycine translocation against the reciprocal of $[Na^+]^2$ — as though the translocated species were $2Na^+$–glycine–X. A Michaelis–Menten-type kinetic analysis showed that the dissociation constant for glycine appeared to decrease as $[Na^+]$ was increased, but that V_{max} was independent of $[Na^+]$. Choline$^+$, tetrakis-(β-hydroxyethyl)-ammonium$^+$, Li^+ and K^+ did not antagonise or substitute for Na^+. By lysing and restoring the red cells in appropriate media[140–142] it was demonstrated that the direction of glycine translocation was determined by the direction of the Na^+ gradient across the membrane[137]. After filling cells with a NaCl–glycine medium and suspending them in the same medium, glycine appeared to be translocated down the Donnan potential gradient (directed outwards) produced by replacing external Cl⁻ by the non-penetrating anion toluene disulphonate[139].

Wheeler et al.[143] have confirmed that for concentrations of Na⁺ above 20 mM, the Na⁺–glycine symporter of pigeon erythrocytes is divalent with respect of Na⁺, but below 20 mM Na⁺, the system is monovalent with respect of Na⁺. They also found that the Na⁺–L-alanine symporter of pigeon erythrocytes and rabbit reticulocytes was monovalent with respect to Na⁺ over a wide range of Na⁺ concentration, and that the Na⁺-linked symporter that translocates α-aminoisobutyric acid in Ehrlich ascites cells is similarly monovalent with respect to Na⁺.

(d) Proton-coupled solute translocation

The classical uncouplers of oxidative phosphorylation, such as DNP and CFCCP*, catalyse the equilibration of the electrochemical activity of protons across natural membranes[144–146]. Horecker et al.[147] observed that DNP catalysed a rapid escape of galactosides through the plasma-membrane of *Escherichia coli* which had previously been induced to accumulate the sugar, and it was accordingly suggested that accumulation of galactosides might be due to the presence of an H⁺–galactoside symporter in the membrane[25]. Using a mutant of *E. coli* possessing a constitutive β-galactoside translocator, but lacking β-galactosidase, Winkler and Wilson[148] showed that when metabolism was inhibited by azide + iodoacetate, lactose and unnatural β-galactosides equilibrated across the membrane *via* a system exhibiting classical circulating-carrier kinetics. Estimations of the effective dissociation coefficients of the carrier centre on either side of the membrane showed that in metabolizing cells the dissociation coefficient on the inside of the membrane was about 100 times greater than on the outside, whereas in non-metabolizing cells the values of the coefficient were presumably equal on either side. On the other hand, the maximum velocity of translocation was the same in metabolizing and non-metabolizing cells. The β-galactoside translocator was unaffected by Na⁺, in agreement with observations of Kolber and Stein[149], but the possibility was not excluded that other ions might be involved. The fact that the specific proton conductor dinitrophenol inhibited galactoside accumulation almost as effectively as azide + iodoacetate, adds further weight to the suggestion that β-galactoside translocation and accumulation in *E. coli* occur *via* an H⁺–β-galactoside symporter.

* CFCCP, carbonylcyanide p-trifluoromethoxyphenylhydrazone.

(i) The ATP/ADP translocator of mitochondria

Studies of the mechanism of inhibition of oxidative phosphorylation by atractyloside in mitochondria led to the discovery[150-154] that ADP and ATP pass between the reversible mitochondrial ATPase and the outer medium *via* an ATP/ADP translocator situated in the cristae membrane. Klingenberg and Pfaff[155] and Heldt[156] described detailed permeability, kinetic and morphological studies carried out by the Marburg group. The ATP/ADP translocator exhibits "saturation" kinetics, has a high temperature coefficient and is very specific towards ADP and ATP. Atractyloside inhibits both entry and exit of ATP and ADP *via* the translocator. The rate of exchange between internal and external nucleotide *via* the ATP/ADP translocator is increased by DNP, but it is not yet known whether the movements of ATP and ADP through the translocator are mutually coupled, or in what state of ionisation or salt formation the nucleotides may be translocated. It is certain, however, that the nucleotides pass across the membrane as complete molecules[155].

(ii) Proton- or hydroxyl ion-coupled anion translocators

Chappell and Crofts[77] and Chappell and Haarhoff[157] have studied the permeability of mitochondria to certain anions and cations by a classical osmotic technique (see refs. 49, 158). They observed that rat-liver mitochondria do not swell in isotonic NH_4^+ chloride, bromide or sulphate or in Na^+ acetate or K^+ acetate, whereas they swell in NH_4^+ phosphate or acetate. It was inferred that the mitochondrial membrane system is impermeable to Na^+ and K^+, but that phosphate and acetate permeate with a neutralising ion, presumably a proton, or in exchange for a neutralising ion, presumably an OH^- ion. It was found that, of the monocarboxylic acids, formate, propionate and butyrate also entered rapidly. The dicarboxylic acids behaved differently. Permeation (swelling) was catalysed by (2 mM) phosphate or arsenate. Phosphate-dependent permeation was observed with succinate, D-malate, L-malate, methylene succinate, malonate and meso-tartrate. Fumarate, maleate, citraconate, mesaconate, D-tartrate or L-tartrate did not permeate under these conditions. The specificity of the permeation reactions for phosphate and arsenate, and amongst the dicarboxylic acids, strongly suggests that substrate-specific translocators exist for phosphate and for the succinate class of dicarboxylic acid.

In the case of the tricarboxylic acids, Chappell and Haarhoff[157] observed that citrate will permeate provided that catalytic amounts of both phosphate (or arsenate) and L-malate are present. This property is shared by *cis*-aconitate, D-tartrate and L-tartrate, but not by *trans*-aconitate, fumarate, maleate, citraconate or mesaconate. No other dicarboxylic acid would replace L-malate. Earlier observations of Gamble[159] on citrate and phosphate translocation can be explained in terms of the translocators identified by Chappell and co-workers.

6. Primary translocation

(a) The Na^+/K^+-antiporter-ATPase

The study of the Na^+, K^+, and Mg^{2+} activated ATPase system, now known to be of wide occurrence in plasma membranes, was initiated by Skou[1] with the object of elucidating the connection between the enzyme-catalysed reactions of metabolism, and the ion-translocation processes. Skou's work has had a most important influence upon the development of knowledge in the biological transport field. It is now generally agreed that the enzyme system that catalyses the Na^+, K^+, Mg^{2+} activated hydrolysis of ATP is the same as, or represents an important component of, the system responsible for the "metabolically driven" extrusion of Na^+ ions and uptake of K^+ ions through the plasma membrane of many types of cells (see refs. 160, 161). The ATPase is particularly active in membrane preparations of excitable and secretory tissues, for example, electric organ[162], brain[163], peripheral nerve[1], muscle[164], salt gland[165], kidney[163], and salivary gland[166]. Many studies have been carried out on red blood cells because of the relative ease of isolating the membrane, and because of the possibility of examining the sidedness of the enzyme system by utilising the phenomenon of reversible lysis[141,142,158,167-169].

The main properties of the Na^+/K^+-antiporter-ATPase have been summarised[36] as follows:

(*1*) The enzyme system is a complex lipoprotein[170,171].

(*2*) Full activity of the ATPase requires the presence of Mg^{2+}, Na^+ and K^+ (refs. 1, 2, 163, 168, 172, 173).

(*3*) The ATPase activity is inhibited by ouabain and by Ca^{2+} (refs. 163, 164, 174).

(*4*) When the ATPase is in an intact membrane, ATP can react with it only from the inside[175], and the ADP produced is also retained on the inside, but it is not clear what happens to the P_i intermediately. In the complete reaction cycle the P_i is probably retained inside[158,164,176].

(*5*) At an early stage of the ATPase reaction cycle, a phosphorylated intermediate $E \sim P$ is formed[2,177,178]. This phosphoryl transfer stage requires Na^+ but not K^+ (refs. 2, 164, 170, 176). The formation of $E \sim P$ is inhibited by Ca^{2+} (refs. 164, 179) but not by ouabain. The Na^+ requirement is directional, and must be satisfied from the inside of the membrane[141].

(6) The hydrolysis of the intermediate, $E \sim P$, requires K^+ (or Rb^+ or Cs^+ or Li^+)[176,180]. The K^+ requirement is directional, and must be satisfied from the outside of the membrane[141].

(7) Ouabain inhibits the hydrolysis of $E \sim P$ and competes with K^+ (refs. 168, 169). The inhibitory effect of ouabain in nerve can be exerted only from outside the membrane[181].

(8) The alkali cations Na^+ and K^+ exert reciprocal competitive effects [3,172,182]. These effects appear mainly to be due to competition of Na^+ for the K^+ site and *vice versa*. Inhibitory effects could also result from the existence of both "in" and "out" sites for Na^+ and K^+.

(9) The phosphorylated intermediate, $E \sim P$, is not a phospholipid, but may be a phosphoprotein[183,184].

(10) In red blood cells it has been estimated that $3Na^+$ pass out and $2K^+$ pass in through the ATPase system per reaction cycle hydrolysing one ATP to one ADP and one P_i[168,172,185].

(11) The kinetics for Na^+ and K^+ activation of the ATPase from rat brain is of the Michaelis–Menten type, but is second order with respect to Na^+ and first order with respect to K^+, indicating that activation involves $2 Na^+$ and $1 K^+$ (refs. 182, 186, 187).

(12) The intact ATPase system of frog-muscle plasma membrane is electrogenic, and the passage of one positive charge outwards per hydrolytic cycle may agree with the membrane potential measurements[188–190].

(13) In red blood cells, the exchange of K^+ across the membrane in the absence of Na^+ but in the presence of glucose appears to require the presence of external P_i (ref. 176).

(14) Some evidence exists for an intermediate formed prior to $E \sim P$ (Ref. 191).

These properties indicate that the active-centre regions of the ATPase are closely related to or are partly or wholly identical to the carrier-centre regions of the Na^+/K^+ translocator.

(b) The H^+-translocator-ATPase

Mitochondria[192], chloroplasts[193–195] and chromatophores[196,197] contain

ATPase systems that are distinct from the Na^+/K^+-antiporter-ATPase of plasma membranes in being insensitive to ouabain[198], activated by Mg^{2+}, but not by $Na^+ + K^+$, and sensitive to oligomycin (ref. 192) or Dio-9 (ref. 199).

The mitochondrial ATPase system has been separated into two major components described by Racker and co-workers as F_0 and F_1. The component F_1 is a protein of molecular weight 280 000 that exhibits a Mg^{2+}-dependent, oligomycin-insensitive ATPase activity[192,200,201]. F_0, on the other hand, is a complex lipoprotein material with no demonstrated relevant enzyme activity[202]. In intact mitochondria, F_1 corresponds to the stalked spheres seen on the inner side of the cristae membrane in electron micrographs[203,204]. Reconstitution of separated F_0 and F_1 gives a material that looks like the cristae membrane with stalked-sphere appendages[205]. If isolated F_1 is recombined with F_0, the ATPase activity becomes oligomycin-sensitive, and it follows that F_0 normally plays a part in the ATPase activity of the F_1–F_0 complex[206]. The fact[156] that ATP is directly accessible to the ATPase system only from inside the intact cristae membrane is consistent with the presence of F_1 on the inner side of the membrane. The relationship appears to be similar in chloroplasts, except that the polarity of the membrane is reversed, the spheres (corresponding to F_1) being on the outside of the grana discs[207].

During ATP hydrolysis, rat-liver mitochondria translocate nearly 2 protons outwards per ATP hydrolysed to $ADP + P_i$ (ref. 208). The ATP-driven proton translocation is sensitive to oligomycin. These and other observations indicate that the mitochondrial ATPase couples the hydrolysis of ATP to the translocation of 2 protons across the membrane in which it resides (see ref. 43).

The fact that ATP can be synthesized by establishing a pH differential of some 3.5 units across the lamellae of spinach chloroplasts[209,210] has been interpreted as evidence in favour of the reversibility of the ATPase system in chloroplasts; and calculations based on the pH differential required to dehydrate $ADP + P_i$ and the stoichiometry of synthesis of ATP per proton estimated to pass through the membrane suggest that the system may be a $2H^+$-translocator-ATPase, as in mitochondria[43]. The reversibility of the mitochondrial ATPase system has similarly been inferred from the recent observations of Reid et al.[211] on the oligomycin-sensitive synthesis of ATP by rat-liver mitochondria subjected to a pH differential.

A fairly detailed mechanism has been proposed for the $2H^+$-translocator-ATPase[42,43].

(c) The H⁺-translocator oxido–reductases

Recent observations on the translocation of protons across the membranes of chloroplasts[212-216], mitochondria[25,146,208,217,218], bacteria[25,26,145], and chromatophores[219] have revived interest in the suggestion of Lundegardh[38] that, by catalysing the translocation of electrons between two half-reactions, the cytochrome system may cause the separate generation of H^+ and OH^- ions, and may thus, effectively catalyse the translocation of protons. Making use of concepts similar to those developed by Davies and Ogston[39], Davies and Krebs[220], Davies[221 222], Conway[40] and Robertson[41], Mitchell has shown that the translocation of protons can be very simply described in terms of the so-called o/r loop[42,43]. The o/r loop consists of two o/r carriers, linked across a membrane so that the terminal oxidant and reductant are in the same phase. The operation of the o/r loop depends upon the currency of reducing equivalents being different in the two arms of the loop, so that protons are produced or taken up at the junction between the two arms of the loop. Various o/r currencies are possible, and if, for example, an o/r loop consisted of one electron carrier and one hydride ion carrier, the stoichiometry of proton translocation would be one proton per $2 \, e^-$ equivalent (loop of type I), whereas if the loop consisted of one hydrogen and one electron carrier the stoichiometry of proton translocation would be two protons per $2 \, e^-$ equivalent (loop of type II).

It has been suggested[43] that the respiratory-chain system of mitochondria may consists of three o/r loops of type II. In support of this suggestion it has been shown that the following stoichiometries for outward proton translocation (written $\rightarrow H^+$) are obtained in suspensions of rat-liver mitochondria: For β-hydroxybutyrate oxidation by O_2, $\rightarrow H^+/O$ approaches 6; for β-hydroxybutyrate oxidation by ferricyanide, $\rightarrow H^+/2 \, e^-$ approaches 4; for succinate oxidation by O_2, $\rightarrow H^+/O$ approaches 4; for succinate oxidation by ferricyanide, $\rightarrow H^+/2 \, e^-$ approaches 2; for ferrocyanide oxidation by O_2, $\rightarrow H^+/O$ is zero[42,43,146,208,218].

Studies on the outward translocation of protons accompanying Ca^{2+} uptake[223-225] are also in accord with a stoichiometry of $2H^+$ translocated outwards per electron pair equivalent traversing each o/r loop. General support for the translocation of protons by the mitochondrial respiratory-chain system was also given by Millard and Robertson[217]. Chance[226], and Rasmussen and Ogata[227] have, however, contested the interpretation of these findings.

7. The coupling of primary and secondary translocation

In primary translocation reactions, chemical particles that are covalently bonded to each other escape in the thermodynamically natural direction. This escape involves the breaking and making of both primary and secondary bonds, and the catalysts of this process in biological systems have long been defined as enzymes. In secondary translocation reactions chemical particles that are not covalently bonded to each other (*i.e.* solutes) escape in the thermodynamically natural direction. This escape involves the breaking and making of secondary and ionic bonds but not of covalent bonds, and the catalysts of this process in biological systems can conveniently be called translocators or porters (avoiding the termination "ase" which signifies the exchange of covalent bonds).

The work done by metabolism arises from the primary bond exchanges, and hence primary translocation reactions are the means whereby the "chemical bond energy" (ref. 27) is transferred to osmotic potential or is expended in the maintenance of osmotic inhomogeneity.

As in the case of coenzyme-linked enzyme reactions[228], primary and secondary translocation reactions are linked when a pair of translocators share a substrate (or cotranslocator). Thus the primary translocation processes, such as ATP-driven Na^+-translocation, or oxido–reduction-driven H^+-translocation may drive secondary translocation processes, such as sugar translocation, or the translocation of Krebs-cycle intermediates. Owing to the electrogenic property of several of the known translocators (*e.g.* the Na^+/K^+-antiporter-ATPase, and the Na^+–glucose symporter), the membrane potential may be involved in the chemical and osmotic balance between coupled translocation reactions. Linkage may occur between pairs (or groups) of secondary translocation reactions that share a solute (*e.g.* Na^+ or H^+). Linkage may also occur between pairs of primary translocation reactions that share a solute.

8. Postscript

It is hoped that the summary of knowledge presented in this chapter under the heading of "active transport and ion accumulation" may incite those interested in this subject not to be content with the rather narrow bounds to which the workers belonging to the so-called "transport union" were formerly accustomed to confine themselves. The term "active transport" is a legacy of an outlook corresponding more or less to the period of biochemical development when the terms "anabolism" and "catabolism" were fashionable. A more complete understanding of the coupling of reactions responsible for the transformation of chemical to osmotic potential robs the concept of active transport of the spectre of the *vis vitae* and enables us to comprehend the active (*i.e.* unnatural) connotation in terms of a thermodynamically natural "down-hill" or passive process. If, in due course, this chapter should be revised, it is to be hoped that the climate of opinion will have changed so that a different title will be thought to be appropriate.

ACKNOWLEDGEMENTS

I would like to record my indebtedness to Dr. Jennifer Moyle for advice and help during the preparation of the manuscript. I am also grateful to Glynn Research Ltd. for supporting this work.

References p. 192

REFERENCES

1   J. C. Skou, *Biochim. Biophys. Acta*, 23 (1957) 394.
2   J. C. Skou, *Biochim. Biophys. Acta*, 42 (1960) 6.
3   J. C. Skou, in A. Kleinzeller and A. Kotyk (Eds.), *Membrane Transport and Metabolism*, Academic Press, New York, 1961, p. 228.
4   R. C. L. Bosworth, *Transport Processes in Applied Chemistry*, Pitman, London, 1956.
5a P. Curie, *J. Phys.*, 3ème Ser. (1894) 393.
5b P. Curie, *Oeuvres*, Gauthier-Villars, Paris, 1908, p. 118.
6   O. Jardetzky and F. M. Snell, *Proc. Natl. Acad. Sci. (U.S.)*, 46 (1960) 616.
7   H. G. Bray and K. White, *Kinetics and Thermodynamics in Biochemistry*, Churchill, London, 1957, p. 106.
8   B. T. Scheer, *Bull. Math. Biophys.*, 20 (1958) 231.
9   O. Jardetzky, *Bull. Math. Biophys.*, 22 (1960) 103.
10 O. Kedem, in A. Kleinzeller and A. Kotyk (Eds.), *Membrane Transport and Metabolism*, Academic Press, New York, 1961, p. 87.
11 O. Jardetzky, *Biochim. Biophys. Acta*, 79 (1964) 631.
12 A. Katchalsky and O. Kedem, *Biophys. J.*, Suppl. 2 (1962) 73.
13 J. R. Moszynski, H. Hoshiko and B. D. Lindley, *Biochim. Biophys. Acta*, 75 (1963) 447.
14 T. Langeland, *Abstr. Commun. Intern. Biophys. Congr., Stockholm*, 1961, p. 56.
15 D. D. Fitts, *Nonequilibrium Thermodynamics*, McGraw-Hill, New York, 1962, p. 35.
16 W. Wilbrandt and T. Rosenberg, *Pharmacol. Rev.*, 13 (1961) 109.
17 H. H. Ussing, *Physiol. Rev.*, 29 (1949) 127.
18 H. H. Ussing, *Advan. Enzymol.*, 13 (1952) 21.
19 B. Anderson and H. H. Ussing, *Acta Physiol. Scand.*, 39 (1957) 228.
20 P. Meares and H. H. Ussing, *Trans. Faraday Soc.*, 55 (1959) 142.
21 P. Meares and H. H. Ussing, *Trans. Faraday Soc.*, 55 (1959) 244.
22 P. Mitchell, in A. Kleinzeller and A. Kotyk (Eds.), *Membrane Transport and Metabolism*, Academic Press, New York, 1961, p. 100.
23 N. S. Bricker, T. Biber and H. H. Ussing, *J. Clin. Invest.*, 42 (1963) 88.
24 P. Mitchell, in A. Kleinzeller and A. Kotyk (Eds.), *Membrane Transport and Metabolism*, Academic Press, New York, 1961, p. 22.
25 P. Mitchell, *Biochem. Soc. Symp., (Cambridge, Engl.)*, 22 (1962) 142.
26 P. Mitchell, *J. Gen. Microbiol.*, 29 (1962) 25.
27 F. Lipmann, *Advan. Enzymol.*, 1 (1941) 99.
28 S. Glasstone, K. J. Laidler and H. Eyring, *The Theory of Rate Processes*, McGraw-Hill, New York, 1941.
29 E. A. Guggenheim, *Modern Thermodynamics by the Methods of Willard Gibbs*, Methuen, London, 1933.
30 F. Lipmann, in E. Nachmansohn (Ed.), *Molecular Biology*, Academic Press, New York, 1960, p. 37.
31 C. A. Vernon, in A. Wassermann (Ed.), *Size and Shape Changes of Contractile Polymers*, Pergamon, Oxford, 1960, p. 109.
32 P. George and R. J. Rutman, *Progr. Biophys. Biophys. Chem.*, 10 (1960) 1.
33 J. B. Wittenberg, *J. Biol. Chem.*, 241 (1966) 104.
34 J. Wyman, *J. Biol. Chem.*, 241 (1966) 115.
35 F. M. Snell, *J. Theoret. Biol.*, 8 (1965) 469.
36 P. Mitchell, *Advan. Enzymol.*, 29 (1967) in the press.
37 J. F. Danielli, *Symp. Soc. Exptl. Biol.*, 8 (1954) 502.

38 H. LUNDEGARDH, *Arkiv Bot.*, 32A, 12 (1945) 1.
39 R. E. DAVIES AND A. G. OGSTON, *Biochem. J.*, 46 (1950) 324.
40 E. J. CONWAY, *Intern. Rev. Cytol.*, 2 (1953) 419.
41 R. N. ROBERTSON, *Biol. Rev. Cambridge Phil. Soc.*, 35 (1960) 231.
42 P. MITCHELL, *Biol. Rev. Cambridge Phil. Soc.*, 41 (1966) 445.
43 P. MITCHELL, *Chemiosmotic Coupling in Oxidative and Photosynthetic Phosphorylation*, Glynn Research, Bodmin, 1966.
44 P. MITCHELL AND J. MOYLE, *Nature*, 182 (1958) 372.
45 P. MITCHELL AND J. MOYLE, *Proc. Roy. Phys. Soc. Edinburgh*, 27 (1958) 61.
46 W. WILBRANDT AND A. KOTYK, *Arch. Exptl. Pathol. Pharmacol.*, 249 (1964) 279.
47 *Enzyme Nomenclature*, Elsevier, Amsterdam, 1965.
48 B. D. DAVIS, *Arch. Biochem. Biophys.*, 78 (1958) 497.
49 H. DAVSON AND J. F. DANIELLI, *The Permeability of Natural Membranes*, 2nd ed., Cambridge University Press, Cambridge, 1952.
50 W. F. WIDDAS, *J. Physiol. (London)*, 118 (1952) 23.
51 W. F. WIDDAS, *J. Physiol. (London)*, 120 (1953) 23P.
52 W. F. WIDDAS, *J. Physiol. (London)*, 125 (1954) 163.
53 T. ROSENBERG AND W. WILBRANDT, *Exptl. Cell Res.*, 9 (1955) 49.
54 L. MICHAELIS AND M. L. MENTEN, *Biochem. Z.*, 49 (1913) 333.
55 H. AKEDO AND H. N. CHRISTENSEN, *J. Biol. Chem.*, 237 (1962) 118.
56 G. A. VIDAVER, *Biochemistry*, 3 (1964) 662.
57 T. ROSENBERG AND W. WILBRANDT, *J. Theoret. Biol.*, 5 (1964) 288.
58 A. C. DAWSON AND W. F. WIDDAS, *J. Physiol. (London)*, 172 (1964) 107.
59 G. G. HAMMES, *Nature*, 204 (1964) 342.
60 H. N. CHRISTENSEN, *Biological Transport*, Benjamin, New York, 1962.
61 W. D. STEIN AND J. F. DANIELLI, *Discussions Faraday Soc.*, 21 (1956) 238.
62 A. K. SOLOMON, in A. KLEINZELLER AND A. KOTYK (Eds.), *Membrane Transport and Metabolism*, Academic Press, New York, 1961, p. 94.
63 P. MITCHELL, in A. KLEINZELLER AND A. KOTYK (Eds.), *Membrane Transport and Metabolism*, Academic Press, New York, 1961, p. 502.
64 B. CHANCE, W. HOLMES, J. HIGGINS AND C. M. CONNELLY, *Nature*, 182 (1958) 1190.
65 F. HOLTON, *Chem. Weekblad*, 53 (1957) 207.
66 H. LUNDEGARDH, *Biochim. Biophys. Acta*, 27 (1958) 653.
67 P. G. LEFEVRE, *Pharmacol. Rev.*, 13 (1961) 39.
68 P. G. LEFEVRE, *Am. J. Physiol.*, 203 (1962) 286.
69 P. G. LEFEVRE, *Biochim. Biophys. Acta*, 120 (1966) 395.
70 C. G. WINTER AND H. N. CHRISTENSEN, *J. Biol. Chem.*, 239 (1964) 872.
71 C. G. WINTER AND H. N. CHRISTENSEN, *J. Biol. Chem.*, 240 (1965) 3594.
72 H. OKADA AND H. O. HALVORSON, *Biochim. Biophys. Acta*, 82 (1964) 538.
73 B. A. NEWTON, *Symp. Soc. Gen. Microbiol.*, 8 (1958) 62.
74 B. C. PRESSMAN, in B. CHANCE (Ed.), *Energy-linked Functions of Mitochondria*, Academic Press, New York, 1963, p. 181.
75 B. C. PRESSMAN, *Proc. Natl. Acad. Sci. (U.S.)*, 53 (1965) 1076.
76 J. B. CHAPPELL AND A. R. CROFTS, *Biochem. J.*, 95 (1965) 393.
77 J. B. CHAPPELL AND A. R. CROFTS, in J. M. TAGER, S. PAPA, E. QUAGLIARIELLO AND E. C. SLATER (Eds.), *Regulation of Metabolic Processes in Mitochondria, (BBA Library, Vol. 7)*, Elsevier, Amsterdam, 1966, p. 293.
78 M. N. SHEMYAKIN, N. A. ALDANOVA, E. I. VINOGRADOVA AND M. U. FEIGINA, *Tetrahedron Letters*, (1963) 1921.
79 R. SARGES AND B. WITKOP, *Biochemistry*, 4 (1965) 2491.
80 E. GROSS AND B. WITKOP, *Biochemistry*, 4 (1965) 2495.

81 S. G. WALEY, *Advan. Protein Chem.*, 21 (1966) 1.
82 H. H. USSING, *Nature*, 160 (1947) 262.
83 C. R. PARK, R. L. POST, C. F. KALMAN, J. H. WRIGHT, L. H. JOHNSON AND H. E. MORGAN, *Ciba Found Colloq. Endocrinol.*, 9 (1956) 240.
84 T. ROSENBERG AND W. WILBRANDT, *J. Gen. Physiol.*, 41 (1957) 289.
85 E. HEINZ, *J. Biol. Chem.*, 225 (1957) 305.
86 E. HEINZ AND P. WALSH, *J. Biol. Chem.*, 233 (1958) 1488.
87 P. MITCHELL, *Symp. Soc. Exptl. Biol.*, 8 (1954) 254.
88 M. LEVINE, D. L. OXENDER AND W. D. STEIN, *Biochim. Biophys. Acta*, 109 (1965) 151.
89 P. MITCHELL, *Nature*, 191 (1961) 144.
90 E. RIKLIS AND J. H. QUASTEL, *Can. J. Biochem. Physiol.*, 36 (1958) 347, 363.
91 T. Z. CSAKY AND M. THALE, *J. Physiol. (London)*, 151 (1960) 59.
92 R. K. CRANE, D. MILLER AND I. BIHLER, in A. KLEINZELLER AND A. KOTYK (Eds.), *Membrane Transport and Metabolism*, Academic Press, New York, 1961, p. 439.
93 D. H. SMYTH, in A. KLEINZELLER AND A. KOTYK (Eds.), *Membrane Transport and Metabolism*, Academic Press, New York, 1961, p. 461.
94 D. H. SMYTH, in A. KLEINZELLER AND A. KOTYK (Eds.), *Membrane Transport and Metabolism*, Academic Press, New York, 1961, p. 488.
95 R. K. CRANE, G. FORSTNER AND A. EICHHOLZ, *Biochim. Biophys. Acta*, 109 (1965) 467.
96 I. LYON AND R. K. CRANE, *Biochim. Biophys. Acta*, 112 (1966) 278.
97 I. BIHLER, K. A. HAWKINS AND R. K. CRANE, *Biochim. Biophys. Acta*, 59 (1962) 94.
98 R. K. CRANE, *Physiol. Rev.*, 40 (1960) 789.
99 F. ALVARADO AND R. K. CRANE, *Biochim. Biophys. Acta*, 56 (1962) 170.
100 F. ALVARADO AND R. K. CRANE, *Biochim. Biophys. Acta*, 93 (1964) 116.
101 S. G. SCHULTZ AND R. ZALUSKY, *J. Gen. Physiol.*, 47 (1964) 567.
102 S. G. SCHULTZ AND R. ZALUSKY, *J. Gen. Physiol.*, 47 (1964) 1043.
103 S. G. SCHULTZ AND R. ZALUSKY, *Nature*, 198 (1963) 894.
104 F. ALVARADO, *Biochim. Biophys. Acta*, 109 (1965) 478.
105 F. ALVARADO, *Biochim. Biophys. Acta*, 112 (1966) 292.
106 T. Z. CSAKY AND U. V. LASSEN, *Biochim. Biophys. Acta*, 82 (1964) 215.
107 P. KOHN, E. D. DAWES AND J. W. DUKE, *Biochim. Biophys. Acta*, 107 (1965) 358.
108 B. R. LANDAU, L. BERNSTEIN AND T. H. WILSON, *Am. J. Physiol.*, 203 (1962) 237.
109 H. NEWEY AND D. H. SMYTH, *Nature*, 202 (1964) 400.
110 J. K. BINGHAM, H. NEWEY AND D. H. SMYTH, *Biochim. Biophys. Acta*, 120 (1966) 314.
111 T. Z. CSAKY, *Ann. Rev. Physiol.*, 27 (1965) 415.
112 T. Z. CSAKY, *Federation Proc.*, 22 (1963) 3.
113 F. ALVARADO AND R. K. CRANE, *Abstr. Biophys. Soc. 10th Meeting*, 1966, p. 130.
114 F. ALVARADO, *Science*, 151 (1966) 1010.
115 A. KOTYK AND A. KLEINZELLER, *Biochim. Biophys. Acta*, 54 (1961) 367.
116 T. CLAUSEN, *Biochim. Biophys. Acta*, 109 (1965) 164.
117 T. CLAUSEN, *Biochim. Biophys. Acta*, 120 (1966) 361.
118 N. KALANT AND R. SCHUCHER, *Can. J. Biochem. Physiol.*, 41 (1963) 849.
119 R. K. CRANE, R. A. FIELD AND C. F. CORI, *J. Biol. Chem.*, 224 (1957) 649.
120 T. R. RIGGS, L. M. WALKER AND H. N. CHRISTENSEN, *J. Biol. Chem.*, 233 (1958) 1479.
121 E. HEINZ, in J. T. HOLDEN (Ed.), *Amino Acid Pools*, Elsevier, Amsterdam, 1962, p. 539.
122 G. TAKAGAKI, S. HIRANO AND Y. NAGATA, *J. Neurochem.*, 4 (1959) 124.

123 P. N. ABADOM AND P. G. SCHOLEFIELD, *Can. J. Biochem. Physiol.*, 40 (1962) 1603.
124 V. G. ALLFREY, J. W. HOPKINS, J. H. FRENSTER AND A. E. MIRSKY, *Ann. N. Y. Acad. Sci.*, 88 (1960) 722.
125 T. Z. CSAKY, *Am. J. Physiol.*, 201 (1961) 999.
126 A. A. YUNIS, G. ARIMURA AND D. M. KIPNIS, *J. Lab. Clin. Med.*, 60 (1962) 1028.
127 H. KROMPHARDT, H. GROBECKER, K. RING AND E. HEINZ, *Biochim. Biophys. Acta*, 74 (1963) 549.
128 M. FOX, S. THIER, L. ROSENBERG AND S. SEGAL, *Biochim. Biophys. Acta*, 79 (1964) 167.
129 H. N. CHRISTENSEN AND M. LIANG, *J. Biol. Chem.*, 240 (1965) 3601.
130 N. BEGIN AND P. G. SCHOLEFIELD, *Biochim. Biophys. Acta*, 104 (1965) 566.
131 N. BEGIN AND P. G. SCHOLEFIELD, *J. Biol. Chem.*, 240 (1965) 332.
132 R. BLASBERG AND A. LAJTHA, *Arch. Biochem. Biophys.*, 112 (1965) 361.
133 B. G. MUNCK, *Biochim. Biophys. Acta*, 120 (1966) 97.
134 H. NEWEY AND D. H. SMYTH, *J. Physiol. (London)*, 170 (1964) 328.
135 D. L. OXENDER AND H. N. CHRISTENSEN, *J. Biol. Chem.*, 238 (1963) 3686.
136 D. L. OXENDER, *J. Biol. Chem.*, 240 (1965) 2976.
137 G. A. VIDAVER, *Biochemistry*, 3 (1964) 795.
138 G. A. VIDAVER, *Biochemistry*, 3 (1964) 799.
139 G. A. VIDAVER, *Biochemistry*, 3 (1964) 803.
140 J. F. HOFFMAN, D. C. TOSTESON AND R. WHITTAM, *Nature*, 185 (1960) 186.
141 R. WHITTAM, *Biochim. J.*, 84 (1962) 110.
142 J. F. HOFFMAN, *J. Gen. Physiol.*, 42 (1962) 9.
143 K. P. WHEELER, Y. INUI, P. F. HOLLENBERG, E. EAVENSON AND H. N. CHRISTENSEN, *Biochim. Biophys. Acta*, 109 (1965) 620.
144 P. MITCHELL, *Biochem., J.*, 81 (1961) 24P.
145 P. MITCHELL, in H. D. BROWN (Ed.), *Cell Interface Reactions*, Scholar's Library, New York, 1963, p. 34.
146 P. MITCHELL AND J. MOYLE, in E. C. SLATER, Z. KANIUGA AND L. WOJTCZAK (Eds.), *Biochemistry of Mitochondria*, Academic Press, London, and P.W.N., Warsaw, 1967, p. 53.
147 B. L. HORECKER, M. J. OSBORN, W. L. MCLELLAN, H. AVIGAD AND C. ASENSIO, in A. KLEINZELLER AND A. KOTYK (Eds.), *Membrane Transport and Metabolism*, Academic Press, New York, 1961, p. 378.
148 H. H. WINKLER AND T. H. WILSON, *J. Biol. Chem.*, 241 (1966) 2200.
149 J. KOLBER AND W. D. STEIN, *Biochem. J.*, 98 (1966) 8P.
150 A. BRUNI, S. LUCIANI AND A. R. CONTESSA, *Nature*, 201 (1964) 1219.
151 A. BRUNI AND G. F. AZZONE, *Biochim. Biophys. Acta*, 93 (1964) 462.
152 E. PFAFF, M. KLINGENBERG AND H. W. HELDT, *Biochim. Biophys. Acta*, 104 (1965) 312.
153 G. BRIERLEY AND R. L. O'BRIEN, *J. Biol. Chem.*, 240 (1965) 4532.
154 J. B. CHAPPELL AND A. R. CROFTS, *Biochem. J.*, 95 (1965) 707.
155 M. KLINGENBERG AND E. PFAFF, in J. M. TAGER, S. PAPA, E. QUAGLIARIELLO AND E. C. SLATER (Eds.), *Regulation of Metabolic Processes in Mitochondria*, (*BBA Library, Vol. 7*), Elsevier, Amsterdam, 1966, p. 181.
156 H. W. HELDT, in J. M. TAGER, S. PAPA, E. QUAGLIARIELLO AND E. C. SLATER (Eds.), *Regulation of Metabolic Processes in Mitochondria*, (*BBA Library, Vol. 7*), Elsevier, Amsterdam, 1966, p. 51.
157 J. B. CHAPPELL AND K. HAARHOFF, in E. C. SLATER, Z. KANIUGA AND L. WOJTCZAK (Eds.), *Biochemistry of Mitochondria*, Academic Press, London, and P.W.N., Warsaw, 1967, p. 75.

158 R. WHITTAM, *Transport and Diffusion in Red Blood Cells*, Arnold, London, 1964.
159 J. L. GAMBLE, *J. Biol. Chem.*, 240 (1965) 2668.
160 J. C. SKOU, *Physiol. Rev.*, 45 (1965) 596.
161 J. D. JUDAH AND K. AHMED, *Biol. Rev. Cambridge Phil. Soc.*, 39 (1964) 160.
162 I. M. GLYNN, *J. Physiol. (London)*, 169 (1963) 452.
163 J. C. SKOU, *Biochim. Biophys. Acta*, 58 (1962) 314.
164 J. JARNEFELT, *Biochim. Biophys. Acta*, 59 (1962) 643.
165 M. R. HOKIN, *Biochim. Biophys. Acta*, 77 (1963) 108.
166 A. SCHWARTZ, A. H. LASETER AND L. KRAINTZ, *J. Cell. Comp. Physiol.*, 62 (1963) 193.
167 G. GARDOS, *Acta Physiol. Hung.*, 6 (1954) 191.
168 R. L. POST, C. R. MERRITT, C. R. KINSOLVING AND C. D. ALBRIGHT, *J. Biol. Chem.*, 235 (1960) 1796.
169 E. T. DUNHAM AND I. M. GLYNN, *J. Physiol. (London)*, 156 (1961) 274.
170 K. AHMED AND J. D. JUDAH, *Biochim. Biophys. Acta*, 93 (1964) 603.
171 R. TANAKA AND K. P. STRICKLAND, *Arch. Biochem. Biophys.*, 111 (1965) 583.
172 A. K. SEN AND R. L. POST, *J. Biol. Chem.*, 239 (1964) 345.
173 T. NAKAO, K. NAGANO, K. ADACHI AND M. NAKAO, *Biochem. Biophys. Res. Commun.*, 13 (1963) 444.
174 P. EMMELOT AND C. J. BOS, *Biochim. Biophys. Acta*, 120 (1966) 369.
175 P. C. CALDWELL, A. L. HODGKIN, R. D. KEYNES AND T. I. SHAW, *J. Physiol. (London)*, 152 (1960) 561.
176 R. L. POST, A. K. SEN AND A. S. ROSENTHAL, *J. Biol. Chem.*, 240 (1965) 1437.
177 J. S. CHARNOCK AND R. L. POST, *Nature*, 199 (1963) 910.
178 J. D. JUDAH, K. AHMED AND A. E. M. McLEAN, *Biochim. Biophys. Acta*, 65 (1962) 472.
179 F. H. EPSTEIN AND R. WHITTAM, *Biochem. J.*, 99 (1966) 232.
180 R. WHITTAM, *Nature*, 196 (1962) 134.
181 P. C. CALDWELL AND R. D. KEYNES, *J. Physiol. (London)*, 148 (1959) 8P.
182 K. AHMED, J. D. JUDAH AND P. C. SCHOLEFIELD, *Biochim. Biophys. Acta*, 120 (1966) 351.
183 M. R. HOKIN AND L. E. HOKIN, *J. Biol. Chem.*, 239 (1964) 2116.
184 I. M. GLYNN, C. W. SLAYMAN, J. EICHBERG AND R. M. C. DAWSON, *Biochem. J.*, 94 (1965) 692.
185 I. M. GLYNN, *J. Physiol. (London)*, 160 (1962) 18P.
186 A. L. GREEN AND C. B. TAYLOR, *Biochem. Biophys. Res. Commun.*, 14 (1964) 118.
187 R. F. SQUIRES, *Biochem. Biophys. Res. Commun.*, 19 (1965) 27.
188 R. H. ADRIAN AND C. L. SLAYMAN, *J. Physiol. (London)*, 175 (1964) 49P.
189 S. B. CROSS, R. D. KEYNES AND R. RYBOVA, *J. Physiol. (London)*, 181 (1965) 865.
190 W. S. CORRIE AND S. L. BONTING, *Biochim. Biophys. Acta*, 120 (1966) 91.
191 A. K. SEN AND R. L. POST, *Abstr. Biophys. Soc. 10th Meeting*, 1966, p. 152.
192 E. RACKER, *Mechanisms in Bioenergetics*, Academic Press, New York, 1965.
193 A. BENNUN AND M. AVRON, *Biochim. Biophys. Acta*, 109 (1965) 117.
194 B. PETRAK, A. CRASTON, F. SHEPPY AND F. FARRON, *J. Biol. Chem.*, 240 (1965) 906.
195 V. K. VAMBUTAS AND E. RACKER, *J. Biol. Chem.*, 240 (1965) 2660.
196 S. K. BOSE AND H. GEST, *Biochim. Biophys. Acta*, 96 (1965) 159.
197 T. HORIO, K. NISHIKAWA, M. KATSUMATA AND J. YAMASHITA, *Biochim. Biophys. Acta*, 94 (1965) 371.
198 D. M. BLOND AND R. WHITTAM, *Biochem. Biophys. Res. Commun.*, 17 (1964) 120.
199 R. E. McCARTY, R. J. GUILLORY AND E. RACKER, *J. Biol. Chem.*, 240 (1965) PC 4822.

200 E. RACKER, in B. CHANCE (Ed.), *Energy-linked Functions of Mitochondria*, Academic Press, New York, 1963, p. 75.
201 H. PENEFSKY AND R. WARNER, *J. Biol. Chem.*, 240 (1965) 4694.
202 Y. KAGAWA AND E. RACKER, *J. Biol. Chem.*, 241 (1966) 2461.
203 E. RACKER, B. CHANCE AND D. F. PARSONS, *Federation Proc.*, 23 (1964) 431.
204 E. RACKER, D. D. TYLER, R. W. ESTABROOK, T. E. CONOVER, D. F. PARSONS AND B. CHANCE, in T. E. KING, H. S. MASON AND M. MORRISON (Eds.), *Oxidases and Related Redox Systems*, Wiley, New York, 1965, p. 1077.
205 Y. KAGAWA AND E. RACKER, *J. Biol. Chem.*, 241 (1966) 2475.
206 Y. KAGAWA AND E. RACKER. *J. Biol. Chem.*, 241 (1966) 2467.
207 D. F. PARSONS, W. D. BONNER AND J. G. VERBOON, *Can. J. Botany*, 43 (1965) 647.
208 P. MITCHELL AND J. MOYLE, *Nature*, 208 (1965) 147.
209 A. T. JAGENDORF AND E. URIBE, *Proc. Natl. Acad. Sci. (U.S.)*, 55 (1966) 170.
210 R. E. McCARTY AND E. RACKER, *Federation Proc.*, 25 (2) (1966) 226.
211 R. A. REID, J. MOYLE AND P. MITCHELL, *Nature*, 212 (1966) 257.
212 A. T. JAGENDORF AND G. HIND, in B. KOK AND A. T. JAGENDORF (Eds.), *Photosynthesis Mechanisms in Green Plants*, Natl. Acad. Sci., Washington, 1963, p. 599.
213 J. NEUMANN AND A. T. JAGENDORF, *Arch. Biochem. Biophys.*, 107 (1964) 109.
214 G. HIND AND A. T. JAGENDORF, *J. Biol. Chem.*, 240 (1965) 3195.
215 G. HIND AND A. T. JAGENDORF, *J. Biol. Chem.*, 240 (1965) 3202.
216 A. T. JAGENDORF AND J. NEUMANN, *J. Biol. Chem.*, 240 (1965) 3210.
217 W. J. T. MILLARD AND R. N. ROBERTSON, *Proc. Natl. Acad. Sci. (U.S.)*, 52 (1964) 996.
218 P. MITCHELL AND J. MOYLE, *Nature*, 208 (1965) 1205.
219 H. BALTSCHEFFSKY AND L. V. VON STEDINGK, in J. B. THOMAS AND J. C. GOEDHEER (Eds.), *Currents in Photosynthesis*, Ad. Donker, Rotterdam, 1966, p. 253.
220 R. E. DAVIES AND H. A. KREBS, *Biochem. Soc. Symp. (Cambridge, Engl.)*, 8 (1952) 77.
221 R. E. DAVIES, in Q. R. MURPHY (Ed.), *Metabolic Aspects of Transport Across Cell Membranes*, University of Wisconsin Press, Madison, 1957, p. 244.
222 R. E. DAVIES, in A. KLEINZELLER AND A. KOTYK (Eds.), *Membrane Transport and Metabolism*, Academic Press, New York, 1961, p. 320.
223 C. ROSSI AND G. F. AZZONE, *Biochim. Biophys. Acta*, 110 (1965) 434.
224 C. ROSSI, A. AZZI AND G. F. AZZONE, *Biochem. J.*, 100 (1966) 4C.
225 C. S. ROSSI, J. BIELAWSKI AND A. L. LEHNINGER, *J. Biol. Chem.*, 241 (1966) 1919.
226 B. CHANCE, in E. C. SLATER, Z. KANIUGA AND L. WOJTCZAK (Eds.), *Biochemistry of Mitochondria*, Academic Press, London, and P.W.N., Warsaw, 1967, p. 93.
227 H. RASMUSSEN AND E. OGATA, *Biochemistry*, 5 (1966) 733.
228 D. E. GREEN, L. H. STICKLAND AND H. L. TARR, *Biochem. J.*, 28 (1934) 1812.

SUBJECT INDEX

Acenes, and carcinogenesis, 52
3-Acetylpyridine–NAD, internal complex, 122
Acetyltryptophan, complex with benzyl-nicotinamide, 121
Acridine complexes, 124, 139
Acridine dyes, mutagenic, and nucleic acids, binding forces, 139
Actinomycin, charge transfer in mode of action, 139
—, reactivity with purines, 35
Active aldehyde, electronic charges in active benzaldehyde, 48
Adenine, complex with trinitrobenzene, 123
—, molecular orbitals, energy coefficients, 31
—, in nucleic acids, formaldehyde action at amino group, 26
—, π- and lone-pair ionization potentials, 25
—, thermodynamic and kinetic stability and role in biochemistry, 22
Adenine–thymine pair, in nucleic acids, electron-donor and -acceptor capacity, role in helix stability, 35
— —, resonance stabilization, 22
Adenosine, complexes, 116, 123
Adenylic acid, complex with chloranil, 123
Alanine, complex with O_2, 125
Alloxan, anionic free radical formation, 34
—, electron-acceptor ability, 34
—, molecular orbitals, energy coefficients, 31
Amethopterine complexes, 126
Amines, donor–acceptor complexes with chloranil or related quinones, conductivities, 151
Amino acid(s), electron distribution, quantum-mechanical calculations, 35, 36

Amino acid(s), *(continuation)*
—, σ-electronic charges and proton shifts, quantum-mechanical calculations, 35, 36
—, proton shifts in NMR spectra, quantum-mechanical calculations, 35, 36
—, quantum-mechanical calculations on electronic distribution and proton shifts in NMR spectra, 35, 36
— translocation, antiport in, 180
— transport, competitive and reciprocal relations between — and sugar transport, 182
Amino acid–Na^+ symporters, 182–184
D-Amino acid oxidase complexes, 120, 121
Aminobenzoic acid complexes, 120, 121
2-Amino-4-hydroxypteridine, molecular orbitals, energy coefficients, 31
2-Amino-4-hydroxypteridine-6-carboxylic acid complexes, 126
Aminopterin complexes, 126
Ammonia complexes, 88
Ammonium ion, hydrogen dibromide salts, 96
—, hydrogen dichloride salts, 96
Aniline, complex with iodine, coordination site, 103
—, —, NMR spectroscopy, 103
Anisole complexes, 88
Anthanthrene, and carcinogenesis, 53
Anthracene, and carcinogenesis, 52
—, complexes, conductivity improvement by light, 161
—, excitation transfer, 72
—, exciton–exciton interaction and photoconductivity, 158
—, exciton generation by ruby laser light, 160
—, fluorescence induction by ruby laser light, 158

Anthracene, *(continuation)*
—, fluorescence sensitization, with traces of naphthacene, 75
—, photoconductivity induction by ruby laser light, 158
—, triplet excitation, spectrum of single-crystal —, 159
Anthrahydroquinone, molecular orbitals, energy coefficients, 31
9,10-Anthraquinone, molecular orbitals, energy coefficients, 31
Anthraquinone sulfonic acid complexes, 110
Anti-coupled solute translocation, (antiport), 175, 180, 181
Antiport, of Na^+ and H^+ across cristae membrane of mitochondria, 181
—, occurrence, 180, 181
Antiporters, natural, 181
Aromatic amines, complex(es), 135
—, — with picric acid, 135
—, and various dyes, lamellar system of complex, photoconductivity, 162
Aromatic hydrocarbons, *see also* Carcinogenic hydrocarbons
—, carcinogenic activity, correlation with electronic structure, 49–53
—, —, K and L regions, 50–53
—, —, localization energies of the K and L regions, 50
—, —, reactivity indices of the K and L regions, 51
—, complex(es) with iodine, conductivities, 151
—, — with trinitrobenzene, energy transfer in, 162
—, polycyclic, FMN complexes and chemical carcinogenesis, 140
—, —, inhibition of flavin-containing enzymes, complex formation, 140
—, —, quinone complexes, and chemical carcinogenesis, 140
Ascorbic acid oxidase, charge-transfer transition and blue colour in presence of oxygen, 141
Atomic orbitals, linear combination of, (LCAO), MO method, 7
ATP, distribution of π-electrons in pyrophosphate chain, 43
—, synthesis, by pH differential across spinach chloroplasts, 188

ATP/ADP translocator of mitochondria, 185
ATPase, H^+-translocator, *see* H^+-translocator–ATPase
—, Na^+/K^+-antiporter, properties, 186, 187
Atractyloside, inhibition of entry and exit of ATP and ADP *via* translocator, 185

Bacteria, H^+-translocator oxidoreductases, 189
Barbituric acid, electron-donor ability, 34
—, molecular orbitals, energy coefficients, 31
1,2-Benzanthracene, and carcinogenesis, 52
Benzene, and carcinogenesis, 52
—, complexes, 88, 89
Benzohydroquinone, molecular orbitals, energy coefficients, 31
Benzoic acid, (substituted), complexes, 111–113, 116, 121, 128
Benzoin condensation, thiamine pyrophosphate catalysis, MO studies, 46, 47
Benzoquinones, (substituted), complexes, 127
p-Benzoquinone, molecular orbitals, energy coefficients, 31
p-Benzoquinone–quinol, *see* Quinhydrone
Benzpyrene complexes, 138, 139
3,4-Benzpyrene, and carcinogenesis, 52
N-Benzyl-3-acetylpyridine, complex with tryptophan, 122
N-Benzyl-3-cyanopyridine, complex with tryptophan, 122
Benzylnicotinamide complexes, 121
(+)-1,1-Binaphthyl, racemization catalysis by organic acceptors, 108
Bioluminescence, 150, 151
Biomolecules, associations considered as CT complexes, 29
—, charge-transfer band, occurrence, 30
—, charge-transfer complexes and electron-donor and -acceptor properties, 27–35
—, conjugated, electron-donor and -acceptor abilities, (table), theoretical evaluation, 31–33
—, —, energy coefficients of MO's of essential groups, 31

Biomolecules, *(continuation)*
—, electron-donor and -acceptor properties, and charge-transfer complexes, 27–35
—, implicated in charge-transfer complexes, 28, 29
Biphenyl complexes, 88
Bond number, definition, 15
Bond order, definition, 15
Brain, mouse, Na^+–amino acid symporters, 182
Bromobenzene complexes, 88
1-Bromobutane complexes, 88
Butadiene, electronic structure, 8, 9
n-Butylamine complexes, 88
Butyryl-CoA dehydrogenase, complex, 121

Caffeine complexes, 111, 123
Calcium ion, inhibition of Na^+/K^+-antiporter–ATPase, 186, 187
Carbonium ions, hydrogen dibromide salts, charge-transfer absorption band, 96
Carcinogenic hydrocarbons, complexes, 127
— —, complexes with purines, role of charge transfer, 138
Carcinogenesis, aromatic hydrocarbons, electronic structure–activity correlation, 49–53
—, bond formation of quinoid type between aromatic hydrocarbons and specific protein fractions, 54
—, chemical, and aromatic hydrocarbon–FMN complexes, 140
—, —, and aromatic hydrocarbon–quinone complexes, 140
—, —, mechanism, electronic aspects, 49–55
—, chemical binding between aromatic hydrocarbons and nucleic acids, 54
—, interactions between carcinogens and cellular receptors, 53–55
—, mechanism of chemical —, electronic aspects, 49–55
—, physical binding between aromatic hydrocarbons and nucleic acids, 54
α-Carotene, molecular orbitals, energy coefficients, 31
β-Carotene, complex with iodine, IR spectroscopy, 102

β-Carotene, *(continuation)*
—, molecular orbitals, energy coefficients, 31
Catechol, complexes, 114
Charge-transfer absorption band(s), in biomolecules, occurrence, 30
— —, characteristics, 89–93
— —, of complexes with same acceptor, ionization potential of donor and position of band, 28
— —, — with 2 distinct maxima, 91
— —, —, and electron affinity of acceptor, 28
— —, — in the infrared, 102
— —, — of iodine with various solvent molecules, (fig.), 128
— —, — and ionization potential of donor, 28
— —, half-maximum height, 90, 91
— —, half-width at half-height, and heat of formation of complex, 91
— —, of indole complexes, 129
— —, transition energy corresponding to, 28
— complexes, *see also* Donor–acceptor complexes
— —, biomolecules implicated in, 28, 29
— —, charge transfer in excited and in ground state, 27
— —, contact, 100, 101
— —, definition, 27, 82
— —, dissociations between biomolecules considered as, 29
— —, electron acceptor, ionization potential of donor and position of CT band, 28
— —, and electron-donor and -acceptor properties of biomolecules, 27–35
— —, in energy transfer in biological systems, 61, 62
— —, forces, ionization potential of donor and degree of association, 28
— — formation, dative-bond wave function, 27
— — —, no-bond wave function, 27
— — —, quantum theory, 27, 28
— — —, and semiconductivity enhancement, 28
— —, between pyridinium salts and different electron donors, 29, 30
— —, in solution, 81–142

Charge transfer, *(continuation)*
—, in the organic solid state, 149–163
—, polymerization initiated by — in complex formation, 98
—, by semiconduction, in the organic solid state, 150–154
— transition(s), in flavoproteins, 131
— —, polarization in coronene–chloranil complex, 93
— —, polarization in quinhydrone, 92, 93
Chemiosmotic processes, thermodynamics, 170–172
Chloranil, complex(es), 85, 93–95, 101, 102, 105, 112, 122, 123, 125, 129, 130
—, — with adenylic acid, 123
—, —, displacement reactions, 95
—, — with nucleic acid bases, nucleosides and nucleotides, 35
o-Chloranil–phthalocyanine film, photoconductivity enhancement of phthalocyanine, 162
Chlorobenzene complexes, 88
Chlorophyll, coated on Cu phenylacetylenide powder, photoconductivity, 163
Chloroplasts, H^+-translocator–ATPase, 187, 188
—, H^+-translocator oxidoreductases, 189
Chlorpromazine, charge transfer in mode of action, 139
—, complexes, 127
Chromatophores, H^+-translocator–ATPase, 187
—, H^+-translocator oxidoreductases, 189
Coenzyme activity, MO studies, 16, 42–49
Coenzymes, mobile electrons responsible for functioning, 42–49
Coenzyme Q, complex with iron, intermediate in oxidative phosphorylation, 141
Conjugated molecules, electronic structure, MO method, 3–15
Copper phenylacetylenide powder coated with chlorophyll, photoconductivity, 163
Copper phthalocyanine, charge-carrier mobility in, 151, 152
—, photoconductivity induction with near IR light, 160
Coronene, complex with chloranil, polarization of charge-transfer transition, 93

Cotransport, biological transport, definition, 175
Crystal energy levels, (fig.), 156
Cyclohexene complexes, 88
p-Cyclophanes, excitation transfer, 71
Cytochrome system, electron transfer by semiconduction, 151
—, electron translocation and metabolic oxido–reduction reactions, 174, 189
—, translocation of protons, 189
Cytochrome b_5 reductase, complex with NAD, 121
Cytosine, molecular orbitals, energy coefficients, 31
—, in nucleic acids, tendency to exist in rare tautomeric form, 23
—, π- and lone-pair molecular ionization potentials, 25

Dative-bond wave function, donor–acceptor complexes, 84
Deamino-NAD, internal complex, 122
Deamino-NADH, internal complex, 122
Dehydroflavin complexes, 102
Delocalization energy, definition, 12
Delocalized electronic system, definition, 8
Deoxycytidine–HCl, complex with chloranil, 123
Deoxycytosine 5′-phosphate, complex with chloranil, 124
Deoxyribonucleic acid, complex with diaminoacridine, 125
2,4-Diamino-6,7-dimethylpteridine complexes, 126
2,4-Diaminopteridine, molecular orbitals, energy coefficients, 31
Diaphragm, Na^+–glucose symporter, 182
1,2,3,4-Dibenzanthracene, and carcinogenesis, 52
1,2,5,6-Dibenzanthracene and carcinogenesis, 52, 54
1,2,7,8-Dibenzanthracene, and carcinogenesis, 52
1,2,7,8-Dibenznaphthacene, and carcinogenesis, 53
1,2,9,10-Dibenznaphthacene, and carcinogenesis, 53
1,2,6,7-Dibenzpyrene, and carcinogenesis, 52
3,4,9,10-Dibenzpyrene, and carcinogenesis, 53

3,5-Dibromotyrosine complexes, 90
Diffusion, facilitated, definition, 175
Diffusion theory, translational facilitated, 173
(+)-9,10-Dihydro-3,4,5,6-dibenzophenan-threne, racemization, catalysis by organic accpetors, 108
Dihydrothioctate complex with FMN, 132
Dihydroxynaphthene, complexes, 114
2,4-Dihydroxypteridine, molecular orbitals, energy coefficients, 31
3,5-Diiodotyrosine complexes, 90, 92
Dimethylamine complexes, 88
N,N-Dimethylaniline, complex with chloranil, formation in polar solvents, 95
Dimethylbenzylsulfonium chloride, solvolysis with phenoxide, complex formation, 107
2,3-Dimethylbutane complexes, 88
Dimethylphenol complexes, 115
Diphenylbutadiene complexes, 89
Diphenylhexadiene complexes, 89
Diphenylhexatriene complexes, 89
1,3-Divinylporphin, molecular orbitals, energy coefficients, 31
DNA, association with aromatic hydrocarbons, role of charge transfer, 138
DNP, catalysis of equilibration of electrochemical activity of protons across natural membranes, 184
—, — of escape of galactoside through plasma membrane of E. coli, 184
—, effect on ATP/ADP translocator in mitochondria, 185
Donor–acceptor complex(es), see also Charge-transfer complexes
—, activation energies for conduction, 151
—, charge-transfer interaction and difference in energy between interacting orbitals, 87
—, chemical considerations for prediction of — formation, 87
—, conductivity improvement by light, 161
—, contact charge transfer, 100, 101
—, coordination sites, identification, use of NMR, 103
—, dative-bond wave function, 84
—, definition, 82
—, energetics, 82–89

Donor-acceptor complex(es),
(continuation)
—, energy of charge-transfer transition and ionization potential of donor, 87, 88
—, energy of the excited state, 85
—, energy of formation, 85
—, energy relationships, diagram, 83
—, equilibrium constants, determination with NMR, 103
—, —, — from shifts in half-way potentials, 104, 105
—, —, — by temperature-jump relaxation method, 106
—, —, higher order complexes, 100
—, —, Scott equation, 99, 100
—, fluorescence, 105
—, formation, and reaction rates, 106–110
—, —, and reduction potential of molecules, 103–105
—, with hexamethylbenzene as donor, free energy of formation, 89
—, infrared spectroscopy, 102
—, intermolecular binding energy, total, 84
—, of iodine, free energy of formation, in relation to ionization potential of donors, (fig.), 88
—, lamellar systems, photoconductivity, 162
—, MO description, 86
—, NMR spectroscopy, 103
—, no-bond wave function, 84
—, paramagnetic, and electron-transfer reactions, 93–99
—, —, ESR signal of radical ions, 93, 94
—, phosphorescence, 105
—, polarography, 103–105
—, quantum yields of phosphorescence and fluorescence, increase of ratio with increase of acceptor concentration, 105
—, racemization of optically active binaphthyl donors, catalysis by organic acceptors, 108
—, resonance interaction, and energy of the excited state, 85, 86
—, —, and energy of the ground state, 85, 86
—, shift in bond-stretching frequencies, 102
—, solvent effects, rules, 101, 102

Donor-acceptor complex(es),
(continuation)
—, stability, 88, 89
—, —, solvent effects, 101
—, with trinitrobenzene as donor, free
    energy of formation, 89
—, valence bond appraoch of Mulliken,
    84–86
Donor–acceptor interaction in enzyme–
    flavin–cofactor binding, 109
Durol, (duroquinol; 2,3,5,6-tetramethyl-
    quinol), 85
—, complex(es), 85, 88, 89
—, — with chloranil, energy of formation,
    85
—, — with tetracyanoethylene, energy of
    formation, 85
—, — with s-trinitrobenzene, energy of
    formation, 85

Ehrlich ascites tumour cells, Na^+–amino
    acids symporter, 182, 183
—, Na^+–glucose symporter, 182
Electron-acceptor abilities of various
    biomolecules, (table, fig.), 31–33
— and -donor properties of biomolecules,
    and charge-transfer complexes, 27–35
Electron-donor abilities of various bio-
    molecules, (table, fig.), 31–33
Electron-transfer, complete, photo-induced
    in donor–acceptor complexes, 93, 94
—, —, thermally induced in donor–
    acceptor complexes, 93
— reactions and paramagnetic complexes,
    93–99
Electron translocation and metabolic
    oxido–reduction reactions in cyto-
    chrome system, 174, 189
Electronic charge around atom, 13
Electronic excitation, propagation in
    biopolymers, 150
— and transfer processes in biological
    dark processes, 62
Energy transfer, charge-transfer complexes
    in, 61, 62
—, mechanisms, 61–77
— by mobile electrons, 61
— through proton transfer, 61
Enzyme activity, MO studies, 42–49
Enzyme cofactor binding, implication of
    donor–acceptor interaction, 109

Erythrocytes of mammals, D-glucose
    translocation, uniport, 179
— —, L-leucine translocation, uniport, 179
— —, Na^+/K^+-antiporter–ATPase, 186,
    187
— of pigeons, glycine translocation, 183, 184
Escherichia coli, H^+–galactoside symporter,
    184
ESR signal in paramagnetic donor–
    acceptor complexes, 93
Ethylamine complexes, 88
Ethylene, electronic structure, 8
Excitation, electronic, *see* Electronic
    excitation
— energy, of aromatic amino acids and
    pigments, 62, 76
— —, electronic, transfer by non-
    radioactive process in biological
    materials, 62
— transfer in biological systems, 76, 77
— —, classifications, 63, 64
— —, in *p*-cyclophanes, 71
— —, in dimeric aggregates of thionine
    and rhodamines, 71
— —, and donor–acceptor distance, 73
— —, excited states involved, 63
— —, experimental evidence, sources, 71
— —, experimental investigations, 71–76
— —, fluorescence sensitization, 72
— —, hopping model, 75
— —, in hydrogen-bonded dimers, 71
— —, interaction mechanisms involved, 64
— —, intermolecular and energy transfer,
    62
— —, long range dipole–dipole transfer,
    68–71
— —, molecular sheet experiments, 73
— —, multistep transfer, 74–76
— —, multistep triplet–triplet transfer,
    75, 76
— —, in polynucleotides, 77
— —, in polypeptides, 77
— —, in pseudo-isocyanine solutions, 71
— —, single-step transfer, 72
— —, single-step triplet–triplet transfer, 74
— —, singlet–higher triplet transfer, 73, 74
— —, strong coupling, absorption spectra,
    65
— —, —, exciton states, 65
— —, —, medium transmitted interaction,
    65

Excitation transfer, *(continuation)*
— —, theory, 63–71
— —, —, application in biological
systems, 16, 17
— —, very weak coupling, 67, 68
— —, —, diffusive propagation, 68
— —, weak coupling, absorption spectra,
Davidov splitting, 66
— —, —, vibronic exciton states, 66
Exciton(s), band, definition, 155, 156
—, charge-transfer, definition, 156
—, definition, 155
— dissociation, by interaction with defect
or impurity in crystals, 158
— —, — with quantum of lattice
vibrational energy, 157, 158
— —, and photoconductivity, 157, 158
—, ionic, definition, 156
— states, excitation transfer, 65
—, Wannier, definition, 156
Exciton–exciton interactions, in
anthracene, photoconductivity, 158
— —, charge-carrier formation in
crystals, 158

FAD, complexes, 112, 116, 117, 120, 121,
135
—, internal complexes, 116
—, intramolecular association of
riboflavin and adenosine moieties,
30
Ferredoxin, *Cl. pasteurianum*, active site
model, semiconduction, 154
Ferric ion, hydrated, charge-transfer
transition, 140
Flash spectroscopy, donor–acceptor
complexes of iodine atoms and
benzenoid hydrocarbons, 106
Flavin coenzymes, complex(es), 109–122,
128–134
— —, — inhibition by polycyclic aromatic
hydrocarbons, 140
— —, — between oxidized and reduced
forms, and pyridine nucleotide
coenzymes, equilibrium relations,
(scheme), 131
—, donor–acceptor interaction in enzyme-
cofactor binding, 109
Flavoproteins, absorption maxima
assigned to charge-transfer transitions,
131

Flavoproteins, *(continuation)*
—, metal-containing, redox catalysis, 142
Fluorescence sensitization, in anthracene
crystals with traces of naphthacene, 75
—, excitation transfer, 72
—, in trypaflavin–rhodamine B system, 72
FMN, complex(es), with dihydrothioctate,
132
—, —, with polycyclic aromatic hydrocarbons, 140
—, —, with serotonin, semiquinone
formation, 113, 129
—, —, with tryptophan, semiquinone
formation, 111, 129, 130
—, in old yellow enzyme, scheme of
Theorell and Nygaard for binding, 109
FMN(H$_2$) complexes, 111–114, 117, 120,
127, 129–133, 140
FMNH$_2$, complex with FMN, charge-
transfer band, 129
—, —, charge-transfer energy estimated
from molecular orbital parameters, 130
—, complex with *N*-methylnicotinamide,
120, 131
—, complex with NAD$^+$, charge-transfer
band, 129
—, —, charge-transfer energy estimated
from MO parameters, 130
Foch equations, MO method, 6
Folic acid, complexes, 126
—, molecular orbitals, energy coefficients,
31
N^{10}-Formylfolic acid complexes, 126
Free valence on atom, definition, 15

Galactoside–H$^+$ symporter in plasma
membranes of *E. coli*, 184
D-Glucose uniporter, in mammalian
erythrocytes, 179
Glucose–Na$^+$ symporter, 181, 182
Glutathione reductase, complex with
NADP, 121
Glycine, complex(es), 113, 125
—, — with O$_2$, 125
— translocation, 182–184
Glycyl-L-tryptophan, complex with
benzylnicotinamide, 121
Gramicidine polypeptides, translocation
of cations, 180
Group-transfer coenzymes, MO studies,
44–49

Guanine, electron-donor ability, 35
—, molecular orbitals, energy coefficients, 31
—, in nucleic acids, alkylation at N-7, 26
—, —, chelation center, 26
—, π- and lone-pair molecular ionization potentials, 25
Guanine–cytosine pair, in nucleic acids, electron-donor and -acceptor capacity, role in helix stability, 35
Guanosine, complex with chloranil, 123
Guanosine–cytosine pair, resonance stabilization, 22
Guanylic acid, complex with chloranil, 124
—, — with diaminoacridine, 125

H$^+$-galactoside symporter in plasma membranes of E. coli, 184
H$^+$-translocator–ATPase, components F_0 and F_1, 188
—, oligomycin effect, 188
—, reversibility, 188
—, sources, 187
H$^+$-translocator oxidoreductases, 189
Haemocyanin, charge-transfer component in absorption spectrum, 141
Hartree–Fock operator, in MO method, 9
Hexamethylbenzene, complex(es), 85, 87–89, 105
—, — with chloranil, energy of formation, 85
—, —, energy of charge-transfer transition, dependence on the first half-wave reduction potential of acceptor, 87
—, — with TCNE, energy of formation, 85
—, — —, formation as function of donor concentration, 105
—, — with tetracyanobenzene, formation as function of donor concentration, 105
—, — with s-trinitrobenzene, energy of formation, 85
—, — —, resonance description, (fig.), 83
—, as donor in complexes, free energy of formation, 89
Histidine, molecular orbitals, energy coefficients, 31
Hückel approximation for conjugated systems, in MO method, 4, 11
Hydrobenzoic acid complexes, 111, 112

Hydrocarbons, aromatic, see Aromatic hydrocarbons
Hydrogen ion, see H$^+$
Hydrogen-bonded dimers, excitation transfer, 71
2-Hydroxy-5-nitrobenzylbromide, reaction with proteins, specificity for tryptophan, 108
N-(β-p-Hydroxyphenylethyl)-3-carbamoyl-pyridinium chloride, absorption and emission spectra, 136, 137
Hypoxanthine, molecular orbitals, energy coefficients, 31

N-(β-4-Imidazolylethyl)-3-carbamoyl-pyridinium chloride, absorption and emission spectra, 136, 137
Indole-2-carboxylate complexes, 121
Indole(s), complex(es), with acetic acid, 121
—, —, charge-transfer bands, 129
—, — with chloranil, charge-transfer band, 129
—, — —, charge-transfer energy estimated from MO parameters, 130
—, — with FMN, in acid solution, spectrum, 133
—, — —, charge-transfer band, 129, 130
—, — —, charge-transfer energy estimated from MO parameters, 130
—, — with NAD$^+$, charge-transfer band, 129
—, — —, charge-transfer energy estimated from MO parameters, 130
—, — —, and analogues, models for enzyme binding, 135
—, (substituted), complexes, 117, 118, 121, 123, 129, 130, 133, 135
Indolylethylnicotinamide, absorption and emission spectra, 136, 137
Infrared spectroscopy, in complex formation studies, 102
Inorganic ion translocation, antiport in, 180
Internal complexes, of NAD (derivatives), 122
Intestinal mucosa, Na$^+$–amino acid symporters, 182, 183
—, Na$^+$–glucose symporter, 181, 182
Iodine, complex(es), 88, 102, 103, 125, 127, 128

Iodine, complex(es), *(continuation)*
—, — with aromatic hydrocarbons, conductivities, 151
—, —, free energy of formation, in relation to ionization potential of donors, (fig.), 88
—, —, with various solvents, charge-transfer bands, (fig.), 128
Ion accumulation and active transport, 167–191
Ionic exciton concept, for aromatic hydrocarbons, theoretical predictions, 157
Iron, complex with coenzyme Q, intermediate in oxidative phosphorylation, 141
Isocyanine, excitation transfer in pseudo-isocyanine solutions, 71

Kidney, Na^+–glucose symporter, 182

Lactam–lactim transformation, in nucleic acids, and resonance energy, 23
Lamellar systems of donor–acceptor complexes, photoconductivity, 162
Laser light, ruby, exciton generation in anthracene crystals, 160
—, —, photoconductivity and fluorescence induction in anthracene, 158
LCAO (linear combination of atomic orbitals), MO method, 7
Leucine, complexes, 113
L-Leucine uniporter, in mammalian erythrocytes, 179, 180
Leucocytes, Na^+–glucose symporter, 182
Lipoprotein membranes, translocation through, kinetic equations, 176–178
Lipoyl dehydrogenase, absorption at 720 $m\mu$, $FADH_2$–NAD^+ complex, 120, 131
—, complexes, 120
Long range dipole–dipole excitation transfer, 68–71
LSD-type drugs, charge transfer in mode of action, 139
—, hallucinogenic activity, and highest occupied MO, 139
Lumichrome, complexes with phenols, 132
Lumiflavin, complexes with phenols, 132

Membranes, lipoprotein, translocation, kinetics, equations, 176–178

Membranes, *(continuation)*
—, plasma, active transport and ion accumulation, 167–191
Menadione, complex(es), 90, 127
—, —, charge-transfer band, fine structure 90
Mercaptide, anaerobic oxidation in presence of nitrobenzene, formation of nitrobenzene anion radical, 97
Mesitylene complexes, 88
Metabolic units, one-carbon, catalysis of transfer by tetrahydrofolic acid, MO studies, 48, 49
Metal cations, complex(es), 140–142
—, — with riboflavin, 141, 142
Methoxybenzoic acid complexes, 111
Methoxyphenylacrylic acid, complex with FMN, 110
p-Methoxyphenyl-1-naphthoyl system on bisteroid framework, excitation transfer, 73
Methylamine complexes, 88
9-Methylfolic acid complexes, 126
N^{10}-Methylfolic acid complexes, 126
1-Methylnaphthalene complexes, 88
N-Methylnicotinamide, complexes, 120, 131
Methylresorcinol complexes, 115
3-Methylriboflavin, complex(es), 110–113
—, — with benzoate, absorption spectrum, 128
—, — with naphthoate, absorption spectrum, (fig.), 128
Mitochondria, cristae membrane, antiport of Na^+ and H^+, 181
—, —, H^+–translocator–ATPase, 188
—, H^+–translocator–ATPase, 187, 188
—, H^+–translocator oxidoreductases, 189
—, Na^+/K^+–antiporter–ATPase, 186
—, o/r loops of respiratory chain, 189
—, outward proton translocation, stoichiometrics for substrate oxidations, 189
—, permeability to anions and cations, 185, 186
—, phosphate-dependent permeation of dicarboxylic acids and tricarboxylic acids, 185, 186
Molecular complex(es), definition, 83
Molecular orbital(s), highest, in the ground state, and first π molecular ionization potential, 12

Molecular orbital(s), *(continuation)*
—, lowest, and electron affinity of molecule, 12
— method, applications of the electronic indices, (table), 14
— —, calculations, present status in biochemistry, (table), 15, 16
— —, electronic structure of conjugated molecules, 3–15
— —, Foch equations, 6
— —, Hartree–Fock operator, 9
— —, Hückel approximation for conjugated systems, 4, 11
— —, LCAO approximation, 7
— —, Pariser–Parr–Pople approximation of SCF method, 10
— —, self-consistent field (SCF) method, 5–7
— —, Wheland–Mulliken approximation, 11
—, π orbitals, definition, 9
—, σ orbitals, definition, 9
Mulliken, valence bond approach, donor–acceptor complexes, 84–86
Muscle cells, skeletal, translocation of Na^+ isotopes, 180
Mutagenesis, and tautomeric shifts in purines and pyrimidines of nucleic acids, 22, 23
Mutation, acridine dyes and nucleic acids, binding forces, 139
—, spontaneous, and G–C into A–T transformation, 23

Na^+–L-alanine symporter, 184
Na^+–amino acid symporters, 182–184
Na^+–glucose symporter, 181, 182
Na^+–glycine symporter, 182, 183
Na^+–imino acid symporter, in rat small intestine, 183
Na^+/H^+ antiport across cristae membrane of mitochondria, 181
Na^+/K^+–antiporter–ATPase, phosphorylated intermediate, 187
—, properties, 186, 187
NAD, binding with enzyme, 135–138
—, (derivatives), internal complexes, 122
NAD(H) complexes, 113, 120–122, 129, 130, 132, 135
NADH (derivatives), internal complexes, 122

NADP, complex with glutathione reductase, 121
NADPH, complexes with FMN, 113
Naphthacene, and carcinogenesis, 52
Naphthalene, and carcinogenesis, 52
—, complexes, 88, 89
— diol, complexes, 133
— —, — with riboflavin, 133
Naphthaquinol complexes, 115, 116
Naphthenic acid (substituted), complexes, 110, 114–116, 128
Naphthohydroquinone, molecular orbitals, energy coefficients, 31
3,4-Naphthopyrene, and carcinogenesis, 53
1,4-Naphthoquinone, molecular orbitals, energy coefficients, 31
Nickel, complex with quinone, charge-transfer bands, 140
Nitrophenols, complexes with FAD, stability, 116, 119, 135
—, (substituted), complexes, 116, 117, 119, 135
NMR spectroscopy, donor–acceptor complexes, 103
No-bond wave function, donor–acceptor complexes, 84
Nuclear magnetic resonance spectroscopy, in donor–acceptor complex studies, 103
Nucleic acids, adenine, formaldehyde action at amino group, 26
—, amino–amine transformation and resonance energy, 23
—, basicity of N atoms, 20
—, charge-transfer forces between stacked base pairs, and structure stability, 24, 27
—, chemical reactivity, quantum-mechanical calculations, 19, 21, 26
—, cytosine in, tendency to exist in rare tautomeric form, 23
—, electron transmission, direct, through the π-systems, 150
—, electronic indices, of nitrogenous bases, and chemical and physico-chemical properties, (figs.), 19, 21
—, electronic structure, 17–27
—, glycosidic linkages, of purines and pyrimidines, susceptibility to hydrolysis, 26
—, guanine, chelation center, 26
—, guanine-alkylating agents, attack at N-7, 26

Nucleic acids, *(continuation)*
—, helix stability, dipole–dipole interaction, 25
—, —, and guanine + cytosine content, 25
—, hydrogen bonds in stability of structure, 20, 22, 23
—, hydrogen bonding and base-pairing specificity, 23, 24
—, lactam–lactim transformations and resonance energy, 23
—, and mutagenic acridine dyes, binding forces, 139
—, nitrogenous bases, chemical and physicochemical properties and electronic indices, (figs.), 19, 21
—, quantum-mechanical calculations, chemical reactivity, 19, 21, 26
—, — —, for "miniature" nucleic acid, 17–19
—, radioresistance and resonance energy per π electron, 22
—, resonance stabilization, G–C pair *versus* A–T pair, 22
—, —, purine *versus* pyrimidine bases, 20
—, semiconductivity, 40, 41
—, tautomeric shifts in purines and pyrimidines and mutagenesis, 22, 23
—, thymine, site of addition reactions, 26
—, Van der Waals–London interactions between base pairs, role in structure stability, 23–26

O_2 complexes, 125
O_2, conduction increase in purines and pyrimidines, 50
Oligomycin, effect on H^+–translocator–ATPase, 188
One-carbon metabolic units, catalysis of transfer by tetrahydrofolic acid, MO studies, 48, 49
Ouabain, effect on H^+–translocator–ATPase, 188
—, inhibition of Na^+/K^+–antiporter–ATPase, 186, 187
Oxidative phosphorylation, coenzyme Q–iron complex as intermediate, 141
— —, inhibition by 1,1,3-tricyano-3-aminopropene, 134

Pancreas, mouse, Na^+–L-proline symporter, 182

Paramagnetic complexes and electron-transfer reactions, 93–99
Pariser–Parr–Pople approximation of SCF MO method, 10
Pentacene, and carcinogenesis, 52
Pentamethylbenzene complexes, 88
Pentaphene, and carcinogenesis, 53
Peptides, electronic transitions in, 39
—, energy levels in, (table), 40
Perylene, excitation transfer, 72
Phenanthrene, and carcinogenesis, 52
—, complexes, 89
Phenols, complex formation with riboflavin, 132, 133
—, (substituted), complexes, 107, 112, 114–116, 132, 133
Phenothiazine, complex(es), 102, 122, 123
—, — with chloranil, infrared spectroscopy, 102
—, electron-donor ability, 33
Phenylacrylic acid, complex(es), 110
—,— with FMN, 110
Phenylalanine, complexes, 113, 125
—, molecular orbitals, energy coefficients, 31
o-Phenylenediamine, complex with chloranil, formation, 95
p-Phenylenediamine, complex with chloranil, infrared spectroscopy, 102
Phenylmethylsulfide, complex with iodine, equilibrium constant, determination with NMR, 103
Phloroglucinol, complex with riboflavin, 115
Phosphates, energy-rich, contributions to energy-wealth, (table), 41, 43
Photoconductivity, 154–161
— in anthracene, ruby laser light, 158
—, charge-carrier production by light energetically insufficient for direct ionization, 155–161
—, in donor–acceptor complexes, 161, 162
—, dye-sensitized, 163
—, and exciton dissociation, 157, 158
—, induction, in Cu phthalocyanine with near IR light, 160
—, in models of lamellar systems of donor–acceptor complexes, 162
Photocurrent and triplet absorption spectrum in anthracene as function of exciting wavelength, 159

Photooxidation, dye-induced, 162
α-Picoline complexes, 88
Picric acid, complex(es), 117, 134, 135
—, — with aromatic amines, 135
—, — with riboflavin, 134
Picryl chloride, catalysis of racemization
    of optically active binaphthyl donors,
    108
$\Delta^{1,2}$-Piperidine-2-carboxylate complexes,
    120, 121
Polarography, in study of donor–
    acceptor complexes, 103–105
Polymerizations, initiated by charge-
    transfer in complex formation, 98
Polynucleotides, excitation transfer in, 77
Polypeptides, excitation transfer in, 77
—, gramicidine, translocation of cations,
    180
Pore uniporters, biological transport, 180
Porphin, molecular orbitals, energy
    coefficients, 31
Porphyrins, metal-containing charge-
    transfer contributions to long-wave-
    length absorption, 141
Porters, biological transport, definition,
    175, 190
Proline, complex with O_2, 125
Proteins, electron transmission, direct,
    through the π-systems, 150
—, π-electronic delocalization, transmit-
    tance across H bonds, 36–38
—, electronic structure, and constituents,
    35–41
—, intrinsic electronic conduction, 150
—, semiconductivity, 35–41
—, —, due to defects or impurities, 40
Pseudo-isocyanine solutions, excitation
    transfer in, 71
Pteridine, molecular orbitals, energy
    coefficients, 31
Purine(s), bases, resonance stabilization,
    20
—, complexes with riboflavin, electrostatic
    forces, 134
—, conduction increase by absorption of
    O_2, 150
—, electron-donor and -acceptor ability, 33
—, interaction with 3,4-benzpyrene, 138,
    139
—, molecular orbitals, energy coefficients,
    31

Purine(s), *(continuation)*
—, in nucleic acids, glycosidic linkages,
    susceptibility to hydrolysis, 26
—, and polycyclic aromatic hydrocarbons,
    increase of solubility, 138
—, reactivity with actinomycin, 35
—, solubilizing power towards aromatic
    hydrocarbons, and electron-donor
    capacity, 35
—, (substituted), complexes, 117–120, 134,
    138, 139
—, tautomeric shifts and mutagenesis, 22,
    23
Pyrene, complex(es), 87, 105, 108
—, — with chloranil, formation as
    function of donor concentration, 105
—, —, energy of charge-transfer transition,
    dependence on the first half-wave
    reduction potential of acceptor, 87
—, — with TCNE, conductivity improve-
    ment by light, 161
—, — with tetracyanobenzene, formation
    as function of donor concentration, 105
—, — with tetracyanoquinondimethide,
    formation as function of donor
    concentration, 105
3-Pyridinealdehyde–deamino-NAD,
    internal complex, 122
2-Pyridinealdehyde–NAD, internal
    complex, 122
Pyridine nucleotide coenzymes, complexes
    between oxidized and reduced forms,
    and flavin coenzymes, equilibrium
    relations, (scheme), 131
Pyridine nucleotide complexes, 135–138
Pyridine, (substituted), complexes, 88, 121,
    122
Pyridinium iodides, charge-transfer bands,
    135
Pyridinium salts, charge-transfer
    complexes with different electron
    donors, 29, 30
Pyridoxal phosphate catalysis, MO
    studies, 44–49
Pyridoxal phosphate, distribution of π-
    electrons in the transitional Schiff
    base, 46
Pyrimidine(s), bases, resonance
    stabilization, 20
—, complexes with riboflavin, electrostatic
    forces, 134

Pyrimidine(s), *(continuation)*
—, conduction increase by absorption of O_2, 150
—, electron-donor and -acceptor ability 33,
—, in nucleic acids, glycoside linkages, susceptibility to hydrolysis, 26
—, (substituted), complexes, 118, 119, 134
—, tautomeric shifts and mutagenesis, 22, 23
Pyrogallol, complex with riboflavin, 115
$\Delta^{1,2}$-Pyrolline-2-carboxylate complexes, 120

Quantum biochemistry, 1–57
Quinhydrone, polarization of charge-transfer transition, 92, 93
Quinol, complex with riboflavin, 114
1,4-Quinol, redox reaction between mono-anion of — and nitrobenzene, 97, 98
Quinones, complexes, 113, 127, 140

Rabbit reticulocytes, Na^+–L-alanine symporter, 184
Radioresistance, of conjugated compounds, and resonance energy per π electron, 22
Resonance energy, definition, 13
Resonance theory, electronic structure of molecules, 3, 4
Resorcinol, complex(es), 114
—, — with riboflavin, 114
Respiratory coenzymes, MO studies, 44
Retinene, molecular orbitals, energy coefficients, 31
Rhodamines, excitation transfer, in dimeric aggregates, 71
Rhodamine B–trypaflavin system, excitation transfer, 72
Riboflavin, absorption above 500 mμ, increase in presence of transition metals, 141
—, complex(es), 102, 106, 110–120, 132–134, 140–142
—, — with Ag ion, charge transfer from Ag ion to flavin, 142
—, — with dihydroflavin, IR spectroscopy, 102, 125
—, —, effect of protonation of isoalloxazine nucleus on acceptor strength of riboflavin, 132, 133
—, — with NADH, of Mahler and Brand, 132

Riboflavin, complex(es) *(continuation)*
—, — with phenols, 132, 133
—, — with purine, electrostatic forces, 134
—, — with pyrimidine, electrostatic forces, 134
—, hydroiodide, complex(es), 132, 133
—, —, — with HI, 95, 96
—, —, — with phenols, 132, 133
—, molecular orbitals, energy coefficients, 31
— semiquinone, charge-transfer chelates, role in redox catalysis in metal-containing flavoproteins, 142
—, shift of long-wavelength maximum in presence of ferrous ion in pyrophosphate buffer, 141

Saccharomyces cerevisiae, α-thioethyl-D-glucopyranoside uniporter, 180
Self-consistent field (SCF) method, MO method, 5–7
Semiconduction, activation energy for conduction in dry proteins and nucleic acids, 150
—, in biology, 149–154
—, in ferredoxin, 154
—, increase by O_2 adsorbed on surface of conducting material, 150
—, in models involving porphyrin-like molecules, system of Wang and Brinigar, 152, 153
—, in polymers, containing chelated iron, system of Dewar and Talati, 152–154
Semiconductivity, enhancement and charge-transfer complex formation, 28
—, in nucleic acids, 40, 41
—, in proteins, 35–41
Semiconductor(s), definition, 149
—, variation of conductivity with temperature, 150
Serotonin, complex(es), 106, 113, 121, 124, 126, 129, 133
—, — with FMN, in acid solution, spectrum, 133
Serotonin creatinine sulfate, complex with riboflavin, rate constants, 106
Single-channel uniporters, biological transport, 180
Singlet exciton, charge-carrier production in the organic solid state, 157–159
Sodium ion, *see* Na^+

Solute translocation, *see* Translocation, solute
Stilbene complexes, 88, 89
Styrene complexes, 88, 89
Sugar translocation, antiport in, 180
Sugar transport, active, 179, 181, 182
—, competitive and reciprocal antiport relations between — and amino acid transport, 182
Sym-coupled solute translocation, 175, 181–184

TCNE, *see* Tetracyanoethylene
TCNQ, *see* Tetracyanoquinondimethide
Temperature-jump relaxation method, determination of rate constants in equilibrium in donor–acceptor complexes, 106
Tetraalkylammonium hydrogen iodide, 96
Tetrachloroquinone complexes, 122, 123, 125
Tetracyanide quinone, complex salts, conductivity, 151
Tetracyanobenzene complexes, 87, 105
Tetracyanoethylene, 85
—, complexes, 85, 91, 105, 123, 124
—, — with substituted benzenes, maxima in charge-transfer band, 91
—, reaction with *N*-alkylanilines, 96, 97
—, — with amines, 96, 97
—, — with indole, 96, 97
Tetracyanoquinondimethide, complexes, 87, 105
—, oxidation of duroquinol, 96
Tetrahydrofolic acid, catalysis of transfer of one-carbon metabolic units, MO studies, 48, 49
Tetrahydrofurane, complex with TCNE, photoresponse on irradiation in charge-transfer band, 162
Tetramethyl-*p*-phenylenediamine, complex(es), 94, 95, 101, 102
—, — with chloranil, complete electron transfer, 94, 95
—, — —, effect of solvent polarity, 101, 102
Tetramethylquinol, (duroquinol, durol), 85
Tetramethyluric acid, complex(es), 138
—, — with pyrene, X-ray crystallography, 138

Thiamine pyrophosphate catalysis of benzoin condensation, MO studies, 46, 47
α-Thioethyl-D-glucopyranoside translocation in *S. cerevisiae*, 180
Thionine, excitation transfer, in dimeric aggregates, 71
Thymidine 5′-phosphate, complex with chloranil, 124
Thymidylic acid, complex with diaminoacridine, 125
Thymine, molecular orbitals, energy coefficients, 31
—, in nucleic acids, site for addition reactions, 26
Toluene complexes, 88, 89
Translational facilitated diffusion theory, 173
Translocases, biological transport, definition, 175
Translocation catalysis, carrier centre, definition, 178, 179
— —, facilitation of solute diffusion by catalytic carriers, 172–174
— —, general mechanisms, 178, 179
— —, through lipoprotein membranes, kinetic equations, 176–178
— —, mobile *versus* fixed carriers, 178
— —, oxygen conduction through solutions of respiratory carriers, 173, 174
—, coupling of primary and secondary, 190
—, electron, and metabolic oxido–reduction reactions in cytochrome system, 174, 189
—, group, definition, 174, 175
—, of protons by cytochrome system, 189
—, proton-coupled solute, 184–186
—, reactions, primary, 175, 186–190
— —, and reaction systems, classification, 175
— —, secondary, 175, 179–186
— —, solute, enzyme-linked, 175, 186–190
—, single-channel, or pore uniporters, 180
—, solute, anti-coupled, 175, 180, 181
—, —, definition, 174
—, —, non-coupled, 175, 179, 180
—, —, sym-coupled, 175, 181–184
—, uniport, 175, 179, 180

Translocation, *(continuation)*
—, uniporter-catalyzed, and enzyme-catalyzed group-transfer processes, 179
Translocator(s), ATP/ADP, of mitochondria, 185
—, biological transport, definition, 175, 190
—, H⁺-, *see* H⁺-translocator
—, proton or hydroxyl-ion coupled ion, 185, 186
Transport, active, and anion accumulation, 167–191
—, —, of sugars, 179, 181, 182
—, biological, *see also* Translocation
—, —, active, definition, 168, 169
—, —, relationships with chemical reactions, 167–170
—,—, translocation catalysis, *see* Translocation, catalysis
—, —, uphill, definition, 169
—, —, vectorial effect of chemical reactions, 169
1,1,3-Tricyano-2-aminopropene, complex(es), 120, 134
—, — with riboflavin, 120
1,1,3-Tricyano-3-aminopropene, complex with riboflavin, 120, 134
—, inhibition of oxidative phosphorylation, 134
Triethylamine complexes, 88
Trimethylamine complexes, 88
Trimethylquinol, complex(es), 133
—, — with riboflavin, 114, 133
Trinitrobenzene, catalysis of racemization of optically active binaphthyl donors, 108
—, complex(es), 83, 85, 89, 122, 123
—, — with adenine, 123
—, —, conductivity improvement by light, 161
—, — with hexamethylbenzene, resonance description, (fig.), 83
—, as donor in complexes, free energy of formation, 89
2,4,7-Trinitrofluorene, catalysis of racemization of optically active binaphthyl donors, 108
2,4,7-Trinitro-9-fluorenyl-*p*-toluenesulfonate, acetolysis in presence of phenanthrene, 107, 108
Trinitrostyrene, copolymerization with vinylpyridines, 98

Triosephosphate dehydrogenase, absorption and donor–acceptor complex formation, 128
Triplet exciton, charge-carrier production in organic solid state, 157–160
Triplet fluorescein, reaction between 2 fluorescein triplets, formation of oxidized and reduced fluorescein, 160
n-Tripropylamine complexes, 88
Trypaflavin–rhodamine B system, excitation transfer, 72
Tryptophan, complex(es), 111, 122, 124–127, 129–131, 134, 135
—, — with chloranil, 130
—, — with FMN, 131
—, —, with NAD⁺, absorption spectrum, (fig.), 135
—, — with riboflavin, 134
—, molecular orbitals, energy coefficients, 31
Tyrosine, complexes, 112, 113
—, molecular orbitals, energy coefficients, 31

Uniport translocation, 175, 179, 180
Uniport-catalyzed translocation, and enzyme-catalyzed group-transfer processes, 179
Uracil, π- and lone-pair molecular ionization potentials, 25
—, molecular orbitals, energy coefficients, 31
Uric acid, antioxidant properties, 34
—, electron-donor ability, 34
—, molecular orbitals, energy coefficients, 31

Valence bond method, electronic structure of molecules, 3, 4
Valinomycin polypeptides, translocation of cations, 180
Vibronic exciton states, excitation transfer, 66
N-Vinylcarbazole, polymerization in presence of electron acceptors, 98
1-Vinyl-5-formylporphin, molecular orbitals, energy coefficients, 31
Vinylpyridines, copolymerization with trinitrostyrene, 98
Vitamin A₁, molecular orbitals, energy coefficients, 31

Vitamin A_2, molecular orbitals, energy
    coefficients, 31
Vitamin K_3, complexes, 90, 127

Wannier exciton, definition, 156
Wheland–Mulliken approximation, MO
    method, 11

Xanthine, molecular orbitals, energy
    coefficients, 31
Xanthopterin complexes, 126
Xylene complex(es), 88, 89

Yohimbine, complex with
    benzylnicotinamide, 121

COMPREHENSIVE BIOCHEMISTRY

Section V – Chemical Biology

Volume 22. Bioenergetics

I. Quantum biochemistry by A. PULLMAN AND B. PULLMAN. II. Mechanisms of energy transfer by TH. FÖRSTER. III. Charge transfer in biology (*a*) Donor–acceptor complexes in solution. (*b*) Transfer of charge in the organic solid state by F. J. BULLOCK. IV. Active transport and ion accumulation by P. MITCHELL. Subject index.

Volume 23. Cytochemistry

I. Nucleus by G. SIEBERT. II. The nucleolus by H. BUSCH. III. Mitochondria by R. J. BARRNETT. IV. Lysosomes by A. L. TAPPEL. V. The cell surface membrane by R. COLEMAN AND J. B. FINEAN. VI. Microbial cytology by M. R. J. SALTON. VII. Endoplasmic reticulum, secretory granules and Golgi apparatus by G. PALADE. Subject index.

Volume 24. Biological information transfer. Viruses. Chemical immunology

I. Biological information transfer. (*a*) DNA synthesis and replication. (*b*) RNA metabolism. 1. Template RNA; 2. Transfer RNA, amino acid activation. (*c*) Protein synthesis. 1. Ribosomal function; 2. The genetic code. (*d*) Biochemical individuality (with general treatment of genetic errors). (*e*) Chemical mutagenesis. (*f*) Phage genetics. II. Biochemistry of viruses. III. Immunochemistry. Subject Index.

Volume 25. Regulatory functions, membrane phenomena

I. Regulatory functions. (*a*) Hormone mechanisms. (*b*) Allosteric effects and feedback mechanisms. (*c*) Enzyme induction and repression. II. Membrane phenomena. (*a*) Bioelectric potentials (incl. nerve impulse). (*b*) Secretion phenomena. (*c*) Cell permeability. Subject index.

Volume 26. Extracellular and supporting structures

I. Biochemistry of the plant cell wall by S. M. SIEGEL. II. Bacterial cell walls by J.-M. GHUYSEN, J. L. STROMINGER AND D. J. TIPPER. III. Somatic and capsular antigens of gram-negative bacteria by O. LÜDERITZ, K. JANN AND R. WHEAT. IV. Chitinous structures by CH. JEUNIAUX. V. Calcified shells by K. M. WILBUR AND K. SIMKISS. VI. Collagen, cartilage and bone by S. FITTON-JACKSON. VII. Dental enamel by M. J. GLIMCHER. VIII. Extracellular fibrous protein: The silks by F. LUCAS AND K. M. RUDALL. IX. Intracellular fibrous proteins and the keratins by K. M. RUDALL. Subject index.

COMPREHENSIVE BIOCHEMISTRY

Section V – Chemical Biology (continued)

Volume 27. Photobiology, ionizing radiations

I. Phototropism by K. V. THIMANN. II. Biochemistry of visual processes by C. D. B. BRIDGES. III. Bioluminescence by F. H. JOHNSON. IV. Photosensitization by M. I. SIMON. V. The effects of ultraviolet radiation and photoreactivation by J. K. SETLOW. VI. Phytochrome and photoperiodism in plants by S. B. HENDRICKS AND H. W. SIEGELMAN. VII. Photosynthesis by L. N. M. DUYSENS AND J. AMESZ. VIII. Effects of ionizing radiations on biological macromolecules by P. ALEXANDER AND J. T. LETT. Subject index.

Volume 28. Morphogenesis, differentiation and development

I. Fertilization by A. MONROY. II. Behaviour of nucleic acids during early development by J. BRACHET. III. Biochemical pathways in embryos by E. SCARANO AND G. AUGUSTI-TOCCO. IV. Factors of embryonic induction by T. YAMADA. V. Biochemistry of amphibian metamorphosis by R. WEBER. VI. Biochemical correlations in insect metamorphosis by L. I. GILBERT. Subject index.

Volume 29. Comparative biochemistry, molecular evolution

I. Comparative biochemistry. (a) Basic concepts. (b) Autotrophic metabolism. (c) Chemical needs in heterotrophs. (d) Biochemical cycles in the biosphere. (e) Biochemistry and taxonomy. II. Molecular evolution. (a) Molecular adaptations to the physical environment. (b) Molecular adaptations to the biological environment. (c) Heteromorphic aspects of molecular evolution. (d) Evolution of biochemical systems, physiological radiations. (e) Biosynthesis and phylogeny. (f) Paleobiochemistry. (g) Chemical evolution and prebiological evolution. Subject index.